CONNECTIONS

CONNECTIONS

James Burke

Little, Brown and Company
Boston · New York · Toronto · London

To Madeline

Revised Edition

Library of Congress Cataloging-in-Publication Data

Burke, James.
    Connections / James Burke. — Rev. ed.
      p.   cm.
    Includes index.
    ISBN 0-316-11672-6
    1. Inventions — History. 2. Technology — History. I. Title.
 T15.B76 1995
 609 — dc20   95-12751

10 9 8 7 6 5 4 3 2 1

HAWK

Design by Robert Updegraff
Picture research by Juliet Brightmore
Artwork research by Dr. Jack Silver
Artwork by Nigel Osborne, Jim Marks, Berry/Fallon
  Design, David Penny, Angus McBride

PRINTED IN THE UNITED STATES OF AMERICA

Contents

Author's Acknowledgements

There are so many people without whose invaluable assistance this book could not have been written – in particular members of university faculties – that it is impossible for me to express my gratitude to each one individually. I hope they will forgive me if I mention only two of their colleagues whose guidance was particularly generous. Professor Lynn White, Jr, of UCLA brought his immense knowledge and wisdom to bear on keeping me on the right track, and Dr Alex Keller of Leicester University was at hand more times than I can remember in moments of panic

I should also like to thank John Lynch for his meticulous and rewarding assistance in research, Mick Jackson and David Kennard for their frequent and sympathetic aid in giving the structure what imaginative expression it has, and the rest of the BBC production team who worked so hard to make possible the television series with which this book is associated: John Dollar, Hilary Henson, Robyn Mendelsohn, Shelagh Sinclair, Diana Stacey, and in particular my assistant, Maralyn Lister.

I should like to compliment Michael Alcock of Macmillan on his unusual ability to make writing a book virtually painless, and to thank Angela Dyer for making order out of shambles, and Robert Updegraff and Juliet Brightmore for investing the text with the kind of illustration worthy of a better work.

Last, but far from least, I thank my long-suffering wife, who has put up with many difficulties during the two years of preparation.

JAMES BURKE
London, 1978

Introduction

In some way or other, each one of us affects the course of history. Because of the extraordinarily serendipitous way change happens, something you do during the course of today may eventually change the world.

As you will see in this book, ordinary people have often made the difference. A self-educated Scottish mechanic once made a minor adjustment to a steam pump and triggered the whole Industrial Revolution. A nineteenth-century weatherman developed a cloud-making device that just happened to reveal to Ernest Rutherford, a physicist he knew, that the atom could be split. Thanks to a guy working on hydraulic pressure in Italian Renaissance water gardens we have the combustion engine. So you don't have to be Einstein to make your mark on events. We all contribute.

This is because there's no grand design to the way history goes. The process does not fall neatly into categories such as those we are taught in school. For example, most of the elements contributing to the historical development of transportation had nothing to do with vehicles. So there are no rules for how to become an influential participant on the web of change. There is no right way. Equally, there is no way to guarantee that your great project meant to alter the course of history will ever succeed.

Things almost never turn out as expected. When the telephone was invented, people thought it would only be used for broadcasting. Radio was intended for use exclusively onboard ships. A few decades ago, the head of IBM said America would never need more than four or five computers.

Change almost always comes as a surprise because things don't happen in straight lines. Connections are made by accident. Second-guessing the result of an occurrence is difficult, because when people or things or ideas come together in new ways, the rules of arithmetic are changed so that one plus one suddenly makes three. This is the fundamental mechanism of innovation, and when it happens the result is always more than the sum of the parts. A silk loom and the 1890 U.S. census gave birth to the computer. Gaslight and the American War of Independence were responsible for raincoats. Glassmaking and English clay made possible transatlantic navigation.

This is the complex process by which the modern world came to have the technological furniture with which it is filled, and which affects all our lives by its very presence. We live surrounded by the end products of thousands of connections. And in every moment that goes by, more connections are made. The world is changing even as you read these words.

This book looks at the forces at work in making the connections that brought into existence some of the most powerful tools and systems that drive the world today: the computer, spacecraft, the production line, television, atomic weapons, plastics, telecommunications, and aircraft. Each of these innovations emerged as the result of a closely linked sequence of events taking place on the great web of history that links us all to each other, to the past, and (in the way that each of us triggers change) to the future.

What tomorrow's connections will be is hard to guess. The great physicist Niels Bohr once said: "Prediction is extremely difficult. Especially about the future." The main difficulty with forecasting the future is that it hasn't yet happened. And however carefully you plan for the future, someone else's actions will inevitably modify the way your plans turn out. However, an understanding of how change happened yesterday may provide clues as to how it might happen again tomorrow. Anyway, there is nowhere else to look for the future but in the past.

Innovation occurs for many reasons, including greed, ambition, conviction, happenstance, acts of nature, mistakes, and desperation. But one force above all seems to facilitate the process. The easier it is to communicate, the faster change happens. Every time there is an improvement in the technology with which ideas and people come together, major change ensues. The Greek alphabet gave birth to philosophy, logic, and the democratic process. The printing press generated the entire Scientific Revolution. The telegraph brought modern business methods into existence and held empires together.

This process has been at work, with extraordinary results, during the brief lifetime of this book. When it was first printed in 1978, there were no laptops, personal digital assistants, electronic agents, Worldwide Webs, commercial on-line services, or cellular phones. That is a measure of the speed and scale with which change is now beginning to happen, as the Information Age advances.

Today, supercomputers and fiber-optic networks, with their ability to make unimaginable amounts of data instantly accessible to millions of people, are accelerating the process of change by many orders of magnitude. Already some scientific disciplines depend on equations that can only be performed by data-processing systems, because they would take a human longer than a lifetime. Daily life is increasingly interactive, and when virtual reality and mobile personal communication numbers are in universal use, the constraints of time and place will disappear and the world will change radically. And at the rate of knowledge-manufacture that will be made possible by the coming technology, a single, life-long specialist qualification will be obsolete, as we scramble to reskill ourselves, perhaps every decade, in order to keep a job. In addition, where we live will no longer mean anything. On the network, we will all live everywhere at once.

The coming surge of change in the very near future will make everything that has happened so far seem as complex as "See Spot Run." Before the tidal wave of innovation breaks upon us, it is vitally important that we understand how change comes about, so as to manage it better, to our general advantage. I hope this book makes a small contribution to that understanding.

JAMES BURKE
London, 1995

1
The Trigger Effect

In the gathering darkness of a cold winter evening on 9 November 1965, just before sixteen minutes and eleven seconds past five o'clock, a small metal cup inside a black rectangular box began slowly to revolve. As it turned, a spindle set in its centre and carrying a tiny arm also rotated, gradually moving the arm closer and closer to a metal contact. Only a handful of people knew of the exact location of the cup, and none of them knew that it had been triggered. At precisely eleven seconds past the minute the two tiny metal projections made contact, and in doing so set in motion a sequence of events that would lead, within twelve minutes, to chaos. During that time life within 80,000 square miles of one of the richest, most highly industrialized, most densely populated areas in the Western world would come to a virtual standstill. Over thirty million people would be affected for periods of from three minutes to thirteen hours. As a result some of them would die. For all of them, life would never be quite the same again.

The moving cup that was to cause havoc unparalleled in the history of North American city life was mounted inside a single back-up electric power relay in the Sir Adam Beck power station at Niagara Falls. It had been set to react to a critical rise in the power flowing out of the station towards the north; the level at which it would trip had been set two years before, and although power levels had risen in the meantime, the relay had not been altered accordingly. So it was that when power on one of the transmission lines leading from Beck to Toronto fluctuated momentarily above 375 megawatts, the magnets inside the rectangular box reacted, causing the cup to begin to rotate. As the spindle arm made contact, a signal was sent to take the overloaded power line out of the system. Immediately, the power it

Manhattan – an island totally dependent on technology. This is 'magic hour', the moment when the skyscraper lights go on just before dusk and the city's energy consumption soars.

had been carrying was rerouted on to the other four northward lines, seriously overloading them. In response to the overload these lines also tripped out, and all power to the north stopped flowing. Only 2.7 seconds after the relay had acted the entire northward output automatically reversed direction, pouring on to the lines going south and east, into New York State and New York City, in a massive surge far exceeding the capacity of these lines to carry it. This event, as the Presidential Report said later, 'occurring at a time of day in which there is maximum need for power in this area of great population density, offered the greatest potential for havoc'.

The first effect was to immobilize the power network throughout almost the entire north-east of America and Canada. Power to heat and light, to communicate and control movement, to run elevators, to operate pumps that move sewage, water and gasoline, to activate electronic machinery is the lifeblood of modern society. Because we demand clean air and unspoiled countryside, the sources of that power are usually sited at some distance from the cities and industries that need it, connected to them by long transmission lines. Due to the complex nature of the way our industrial communities operate, different areas demand power at different times; for this reason the transmission lines operate as a giant network fed by many generating stations, each one either providing spare power or drawing on extra, according to the needs of the particular area. As a result of this net- work, failure in one area can mean failure in all areas. The generators producing power for the transmission lines can be run at various speeds, which determine the frequency of the current; in order that all the inputs to a network may work together they must all be set to produce the same frequency, so the generators must run at whatever speed their design demands to produce that frequency. This maintains what is called a 'stable' system. When something goes wrong, such as the massive overload of 9 November, the system becomes wildly unstable. Protective devices automatically cut in, to protect indivi- dual generating stations from the overload by isolating them from the network. This means that they will then be producing either too much power for the local area, or too little. The sudden change in speed on the part of the generators to match output to the new conditions can cause serious damage, and for this reason the generators must be shut down.

This is what happened throughout most of north-east America in the twelve minutes following the relay operation at Beck. Over almost all of New York City the lights flickered and went out. Power stopped flowing to the city's services. An estimated 800,000 people were trapped in the subways. Of the 150 hospitals affected, only half had auxiliary power available. The 250 flights coming into John F.

Kennedy airport had to be diverted; one of them was on its final approach to landing when the lights on the runway went out, and all communication with the control tower ceased. Elevators stopped, water supplies dried up, and massive traffic jams choked the streets as the traffic lights stopped working. All street lighting went out in a city of over eight million inhabitants. To those involved the event proved beyond shadow of doubt the extent to which our advanced society is dependent on technology. The power transmission network that failed that night is a perfect example of the interdependent nature of such technology: one small malfunction can cripple the entire system.

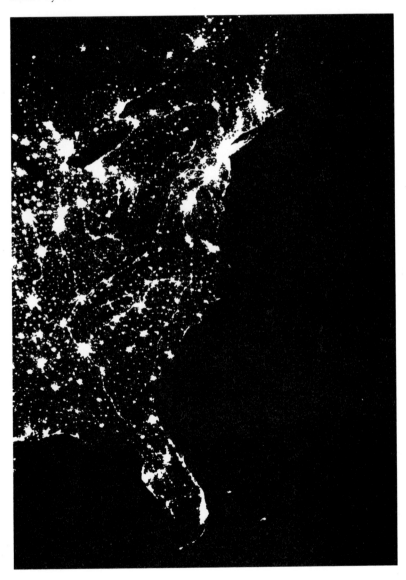

In this satellite photograph of the eastern seaboard of the United States, taken at night, the cities blaze with light, from Boston (top right) to Miami (bottom right). The area affected in the 1965 blackout ran north from New Jersey, and included the most highly illuminated concentration of population in the photograph, New York and Boston.

This interdependence is typical of almost every aspect of life in the modern world. We live surrounded by objects and systems that we take for granted, but which profoundly affect the way we behave, think, work, play, and in general conduct our lives and those of our children. Look, for example, at the place in which you are reading this book now, and see how much of what surrounds you is understandable, how much of it you could either build yourself or repair should it cease to function. When we start the car, or press the button in an elevator, or buy food in a supermarket, we give no thought to the complex devices and systems that make the car move, or the elevator rise, or the food appear on the shelves. Today we are almost totally dependent on the products of science and technology. They have already changed our lives: at the simplest level, the availability of transport has made us physically less fit than our ancestors. Many people are alive only because they have been given immunity to disease through drugs. The vast majority of the world's population relies on the ability of technology to provide and transport food.

The prime example of man's love–hate relationship with technology: the motor car, which makes mobility possible, and the traffic jam, which makes it impossible. Will freeways remain, long after the motor car has become obsolete, as monuments to the ability of technology to alter the shape of the world around us?

There is enough food only because of the use of fertilizers. The working day is structured by the demands of the mass-production system. Roads are built to take peak hour traffic and remain half-empty outside those hours. We can neither feed, nor clothe, nor keep ourselves warm without technology.

The objects and systems produced by technology to perform these services operate interdependently and impersonally. A mechanical failure or industrial unrest in a factory that makes only one component of an automobile will affect the working life of thousands of other people working in different factories on other components of the same car. Step across the road into the path of an oncoming vehicle and your life may depend on the accuracy with which the brakes were fitted by someone you do not know and will never meet. A frost in Brazil may change your coffee-drinking habits by making the price prohibitive. A change of policy in a country you have never visited and with which you have no personal connections may radically alter your life – as was the case when the oil-producing states raised the price of oil in 1973 and thus set off rampant inflation throughout the Western world. Where once we lived isolated and secure, leading our own limited lives whose forms were shaped and controlled by elements with which we were intimately acquainted, we are now vulnerable to change which is beyond our own experience and control. Thanks to technology no man is an island.

Paradoxically this drawing-together of the community results in the increasing isolation of the individual. As the technological support systems which underpin our existence become more complex and less understandable, each of us feels less involved in their operation, less comprehending of their function, less confident of being able to operate without them. And although international airlines criss-cross the sky carrying millions of passengers every day, only a tiny fraction of the world's population has ever flown, let alone visited a foreign country or learned a foreign language. We gain our experience of the world from television. The majority of the people in the advanced industrialized nations spend more time watching television than doing anything else besides work. We plug in to the outside world, enjoying it vicariously. We live with the modern myth that telecommunications have made the world smaller, when in reality they have made it immeasurably bigger. Television destroys our comfortable precon-ceptions by showing us just enough to prove them wrong, but not enough to replace them with the certainty of first-hand experience. We are afforded glimpses of people and places and customs as and when they become newsworthy – after which they disappear, leaving us with an uncomfortable awareness that we know too little about them.

In the face of all this most of us take the only available course: we ignore the vulnerability of our position, since we have no choice but to do so. We seek security in the routines imposed by the technological systems which structure our lives into periods of work and rest. In spite of the fact that any breakdown in our interdependent world will spread like ripples in a pool, we do not believe that the breakdown will occur. Even when it does, as in New York in 1965, our first reaction is to presume that the fault will be rectified, and that technology will, as it always has, come to the rescue. The reaction of most of the New Yorkers trapped in subways, elevators, or unlit apartment blocks was to reach out to the people immediately around them – not to organize their own escape from the trap, but to share what little warmth or food they had so as to pass the time until danger was over. To have considered the possibility that the failure was more than a momentary one would have been unthinkable. As one of the sociologists who studied the event wrote: 'We can only conclude that it is too much to ask of us poor twentieth-century humans to think, to believe, to grasp the possibility that the system might fail . . . we cannot *grasp* the simple and elementary fact that this technology can blow a fuse.' The modern city-dweller cannot permit himself to think that his ability to cope in such a situation is in doubt. If he did so he would be forced to accept the uncertainty of his position, because once the meagre reserves of food and light and warmth have been exhausted, what then?

At this point another myth arises: that of the escape to a simpler life. This alternative was seriously considered by many people in the developed countries immediately after the rise in oil prices in 1973, and is reflected in the attitudes of the writers of doomsday fiction. The theory is that when sabotage or massive system failure one day ensures the more or less permanent disruption of the power supply, we should return to individual self-sufficiency and the agrarian way of life. But consider the realities of such a proposal. When does the city peasant decide that his garden (should he possess one) can no longer produce enough vegetables (should he know how to grow them and have obtained the necessary seeds and fertilizer) and animal protein and fats (should he know where to buy an animal and rear it) to support him and his family? At this stage, does he join (or worse, follow) the millions who have left the city because their supplies have run out? Since the alternative is to starve, he has no choice.

He decides to leave the city. Supposing he has the means of transport, is there any fuel available? Does he possess the equipment necessary for survival on the journey? Does he even know what that equipment is? Once the decision to leave has been taken, the modern city-dweller is alone as he has never been in his life. His survival is, for the first time, in his own hands. On the point of departure, does he know in which direction to go? Few people have more than a hazy notion of the agriculturally productive areas of their own country. He decides, on the basis of schoolbook knowledge, to head for one of these valleys of plenty. Can he continue to top up his fuel tanks for as long as it takes to get there? As he joins the millions driving or riding or walking down the same roads, does he possess things those other refugees might need? If so, and they decide to relieve him of them, can he protect himself? Assuming that by some miracle the refugee finds himself ahead of the mob, with the countryside stretching empty and inviting before him, who owns it? How does he decide where to settle? What does a fertile, life-sustaining piece of land look like? Are there animals, and if not, where are they? How does he find protection for himself and his family from the wind and rain? If shelter is to be a farmstead – has it been abandoned? If it has not, will the occupier be persuaded to make room for the newcomers, or leave? If he cannot be so persuaded, will the refugee use force, and if necessary, kill? Supposing that all these difficulties have been successfully overcome – how does he run a farm which will have been heavily dependent on fuel or electricity?

Of the multitude of problems lying in wait at this farm, one is paramount: can the refugee plough? Plants will grow sufficiently regularly only if they are sown in ploughed ground. Without this talent – and how many city-dwellers have it? – the refugee is lost: unless he has a store of preserved food he and his family will not survive the winter. It is the plough, the basic tool which most of us can no longer use, which ironically may be said to have landed us in our present situation. If, as this book will attempt to show, every innovation acts as a trigger of change, the plough is the first major man-made trigger in history, ultimately responsible for almost every innovation that followed. And the plough itself came as a result of a change in the weather.

At the end of the last ice age, in about 10,000 B.C., the glaciers began to retreat and the summer temperature began to rise. With the increase in temperature came a diminution of rainfall. This climatic change was disastrous for the hunting nomads living in the high grass-lands: vegetation began to dry up and disappear, taking with it the herds on whose survival the nomads depended. Water became scarcer, and eventually, some time between 6000 and 5000 B.C., the hunters came down from the plateaux in search of regular food and water. They came down in northern India, central America, Syria and Egypt; they may also have come down in Peru. They came initially looking for animals which had themselves gone in search of water, and found it in river valleys. One of these valleys was uniquely suited to the development of a unified community: the Nile.

The nomads who, for example, descended to the rivers Tigris and Euphrates in Syria spread out and eventually formed themselves into individual city states. But in Egypt the new settlers found a fertile ribbon of land 750 miles long, limited on either side by inhospitable scrubland and desert; they were united by the continuity of the great river common to all. There was nowhere else to go. 'Egypt', said the Greek historian Herodotus, 'is the gift of the Nile'. Initially the gift was of animals, sheltering in and living off the reeds and marshland along the edge of the river, and including birds, fish, sheep, antelope, wild oxen and game animals. As the settlers erected their primitive shelters against wind and rain and attempted to domesticate the animals – there are even records in their tombs of attempts to tame hyenas and cranes – someone may have noticed the accidental scattering by the wind of seeds on newly watered ground at the river's edge, and the growth of new plants that followed. This action must have been imitated successfully, because at some time around 5000 B.C. the nomads decided not to move on as before, but to remain permanently. This decision can only have been made because of sufficient food reserves. The gift of the Nile was now something different from animals: it was fertility of soil.

The Nile itself is formed of two rivers, the White Nile, rising in the African lakes far to the south, and the Blue Nile, which falls from the Abyssinian plateau. One brings with it decaying vegetable matter from the lakes, and the other carries soil rich in potash from the plateau. This is a perfect mixture for fertilizing the ground. The land on the edge of the Nile had no need of manure. It would, with the minimum of tillage, produce full crops of emmer wheat and barley.

The earliest agricultural tool, the digging stick, illustrated on the wall of the tomb of Nakht, in Thebes. The stick made holes for the seeds, which were later trodden in by asses. This tool was also used in early construction of irrigation ditches, vital to the Egyptian economy, and became one of the symbols of Pharaonic power.

Initially tillage was probably done by hand, the farmer merely breaking open the ground and laying the seeds in separate holes. But as this produced more grain to feed more mouths, the population must have increased to the point where such haphazard methods were insufficient, and the next step was taken. Pointed digging sticks pulled by hand would open the ground faster. The decisive event occurred some time before 4000 B.C. when it was noticed, after centuries of experience, that the river rose and fell regularly once a year, and that within the year there were three distinct periods: the flood, the retreat of the waters, and the drought. With this early understanding of the annual nature of the flood came an awareness of the need to match human agricultural activity to the cycle, and to harness the retreating waters for use during the dry period. Primitive mud ridges were thrown up to trap the water in basins, and ditches were dug from the basins to the fields to carry water to the growing crops. At about the same time as these first attempts at irrigation, the digging stick changed its shape; it became a simple scratch plough, with a forward-curving wooden blade for cutting the soil, and a backward-curving pair of handles with which the farmer could direct the oxen which now replaced men as a source of traction power. This simple implement may arguably be called the most fundamental invention in the history of man, and the innovation that brought civilization into being, because it was the instrument of surplus.

The trigger of civilization, the scratch plough. This small wooden model from an Egyptian tomb is dated around 2000 B.C. In nearby Mesopotamia, where the plough also appeared early, the principal crop was barley – which may account for the reputation the Mesopotamians had for alcoholism.

A community may continue to exist as long as it has adequate food, and it may expand as long as food production can keep up with the increase in numbers, but it is not until it can produce food which is surplus to requirements, and is therefore capable of supporting those who are not food producers, that it will flourish. This development was made possible by the plough, and it caused a radical transformation of Egyptian society. The earliest evidence of the political structuring of the country that followed is shown in remains dating from before 3000 B.C. which show the division of the land into 'water provinces', rectangular shapes criss-crossed by lines indicating canals. The governors of these areas were called *adj mer* ('diggers of canals'), and it was their duty to ensure a regular supply of water to the fields. One of the earliest relics of the kings of the time is a mace head, carved with pictures showing the king carrying a hoe with which to build or open canal walls and water basins.

Within a brief period of these early beginnings, Egypt had developed a sophisticated centralized civilization. Initially the surplus produced by irrigation and ploughing permitted non-foodproducers to operate within a community, and in the beginning these may have been the men who dug and maintained the irrigation systems, and those who organized them. These administrators would have derived their authority from the knowledge of astronomy which gave them alone the magic ability to say when the flood would come, when to sow on the land wet from the receding waters, and when to harvest. The grain needed storage room out of the weather, and dried clay daubed on woven reed baskets gradually gave way to more permanent containers as the demand for them increased with the crop. Fire-hardened clay pots, made from spiralling loops of wet earth, came to be used, and the first evidence of solid building is of piles of these pots heaped together to form a central granary. The need to identify the ownership and amount of grain contained in a pot or a granary led to the development of writing. The first picture-words come from before 3000 B.C. and comprise lists of objects and totals of figures contained in pots and chests. The surplus grain paid for craftsmen: carpenters, potters, weavers, bakers, musicians, leather-workers, metal-workers, and those whose task it was to record everything – the scribes. The need to ensure regularity of harvest in order to support these members of the community demanded a taxation system, and so that it should be operated fairly skills were developed to assess each man's due. Initially this may have started with the measurement of field boundaries destroyed each year by the flood, but as time passed and the irrigation systems grew more complex, the process demanded greater sophistication, calling for mathematics to handle the measurement of distance, area and cubic amount.

These early forms of arithmetic and geometry grew from the demands of canal building: how long, how wide and how deep? It may have been the need for tools to do the job which spurred interest in the copper deposits across the Red Sea in Sinai, and this in turn would have stimulated the use of metal for weapons. Weapons were needed by those whose task it was to protect the land and crops from invasion, as the surplus food and the goods financed by its production began to be used as barter with neighbouring communities, some of which looked with envy on the riches of Egypt. Metal tools gave the Egyptians the ability to work stone, initially, perhaps, in blocks for strengthening the irrigation ditches. The Nile is bordered for 500 miles south from Cairo by limestone cliffs, and it is from this stone that the first pyramid was constructed.

A mere hundred and fifty years after the first use of stone for the construction of buildings, the massive step pyramid of King Djoser was erected. It rises out of the desert at Saqqara, south of Cairo. Built by the king's chief minister, Imhotep, it is the oldest extant stone structure in the world, dating from around 2800 B.C. It was constructed using the tools and the theoretical knowledge developed by the canal builders, and it shows a high degree of precision in the use of both. By the time Djoser was being laid in his pyramid, Egyptian society had developed a form that is little changed today. At the top came the Head of State, served by his cabinet of advisers; these were

This magnificent wall painting is in the tomb of the vizier Rekhmire, in Thebes. It shows shoemakers at work (top row) and carpenters (middle row) using bow drills, saws, adzes and chisels. Note how, in the absence of a plane, three men are smoothing a beam with rubbing stones. Bottom right, a metal-worker uses a blowpipe to raise the temperature of his fire.

aided by a civil service which organized every aspect of life in the state, gathering taxes from craftsmen and farmers to support themselves and the army. The regulation of the state's business was effected through the application of laws, which rested for their observance on the availability of an annual calendar, by now divided into twelve months of thirty days each. By 2500 B.C. the Egyptians (and their neighbours the Mesopotamians) had a developed and sophisticated society operating with a handful of essential tools: civil engineering, astronomical measurement, water-lifting machinery, writing and mathematics, primitive metallurgy, and the wheel. With these tools the Egyptians administered an empire whose power and influence was unparalleled in the ancient world, based on an agricultural output made possible by the plough. Its use had ensured the continued survival and expansion of the community and set in motion the changes that resulted from that expansion

The first man-made harvest freed mankind from total and passive dependence on the vagaries of nature, and at the same time tied him for ever to the very tools that set him free. The modern world in which we live is the product of that original achievement, because just as the plough served to trigger change in the community in which it appeared, each change that followed led to further change in a continuing sequence of connected events.

Part of a bas-relief from the tomb of Menna (c. 1400 B.C.) shows the early development of geometry and mathematics, used in the measuring of fields for taxation and in calculating amounts of grain at harvest time. The scribes (lower right) are holding ink palettes and writing on papyrus with reed pens.

The author at the step pyramid of King Djoser. The basic structure was of rubble and stone, faced with limestone slabs. Since the ancient Egyptians did not have pulleys the standard method of construction was to haul the stone up sloping ramps of earth which were removed when the pyramid was complete.

The story of the direction taken by that sequence of events is the subject of this book. The reason why each event took place where and when it did is a fascinating mixture of accident, climatic change, genius, craftsmanship, careful observation, ambition, greed, war, religious belief, deceit, and a hundred other factors. Following the trail of events that leads from some point in the past to the emergence of a modern invention that affects our lives is like being involved in a detective story, in which the reader will know at any particular stage in the story's development only as much as did the people of the time. As each story unfolds it will become clear that history is not, as we are so often led to believe, a matter of great men and lonely geniuses pointing the way to the future from their ivory towers. At some point every member of society is involved in the process by which innovation and change comes about, and this book may help to show that given average intelligence and the information available to the innovators of the past, any reader could have matched their achievements.

The clue to the trigger which sets off the first of these detective stories is this: how did a modern invention whose existence threatens the life of every human being on Earth start 2600 years ago with a discovery made in a river in Turkey?

2
The Road
from Alexandria

In north-western Turkey, not far inland from the Aegean Sea, stands a mountain known in ancient times as Mount Tmolus, from which two rivers fall, one the Pactolus, the other the Hermus. Several factors combine to make these two rivers important in history. The gradient they follow is shallow, so they flow slowly; as they cross the coastal plain they fan wide and smooth, carrying loads of deposit brought down from Tmolus in the form of fine grains of soil mixed with gold, which are winnowed from the slopes of the mountain by the action of wind and temperature change, and washed away by the waters. According to the Greek historian Herodotus, writing in the fourth century B.C., these two rivers had for centuries been known as the richest source of panned gold in the world. Two thousand seven hundred years ago they lay within the boundaries of the kingdom of Lydia, and in one of them some unknown prospector, probably in the employ of the king, made one of the most fundamental discoveries in the history of mankind.

At the time, the standard method for retrieving and smelting the gold dust was through the use of sheepskins. The grease in the skins would trap the tiny particles of gold, and when the skin was fully laden it would be hung on a branch to dry, then thrown into a furnace where the heat would incinerate the animal material, leaving the gold lying in blobs among the fine ashes. It may have been this use of the sheepskin that gave rise to the myth of the Golden Fleece sought by Jason and his Argonauts. Be that as it may, the gold blobs were melted into blocks, or ingots; in this form they were used as a replacement for goods and as payment for services – a practice which goes back as far as the third millennium B.C. in Mesopotamia and Egypt, where the

On 21 November 1783, in Paris. The trip carried two men – Pilatre de Rogier and the Marquis d'Arlandes – five miles from their lift-off in the Bois de Boulogne and lasted twenty-five minutes. The balloon was to prove invaluable in the early years of research into the behaviour of the upper atmosphere, providing much of the basic data on which the modern science of meteorology is founded.

value of the ingot of gold (or copper, or silver) was determined by weight. The limitations of such a system are obvious. The ingots were bulky and difficult to transport. They could only be used for payment on a large scale such as took place between one state and another, or for settling accounts with mercenaries at the end of their period of service.

The discovery made by the man panning for gold in the Pactolus changed things at a stroke. Apart from fine grains of gold, the river also contains small, flat pieces of a flinty stone, black in colour, whose proper geological name is schist. The first reference to the use of schist is by Herodotus, who says that the Lydians cut the top surface of the stone flat, leaving it matt. If gold were rubbed on this matt surface it would make scratch marks. Pure gold would leave yellow marks, gold mixed with silver, white ones, and gold mixed with copper, red marks. The stone could thus be used to assay the quality of the gold, and its common name has passed into our language as a metaphor for evaluation: the touchstone.

The effect of this accidental discovery, made some time in the eighth century B.C., was to be immense. It gave the rulers of Lydia, probably starting with King Gyges (685 B.C.), the ability to ensure a standard quality to their precious metal, for the touchstone shows to even the most inexperienced eye a difference in quality of the smallest percentage. It had been the custom for centuries in the Babylonian and Egyptian empires to stamp ingots with some mark giving authority to

A fifth-century B.C. bowl depicting Jason and the dragon guarding the Golden Fleece. The myth may well spring from events relating to a raid on a community in order to obtain their more advanced knowledge of metallurgy.

A sixth-century B.C. Lydian stater, *made of a gold–silver alloy called electrum and punch-marked by the issuing mint.*

Portraits first appeared regularly on coins after Alexander's death. This silver tetradrachm, dated about 290 B.C., shows Alexander as a god with the Zeus Ammon ram's horn growing from his head. It was minted in Pergamum for Lysimachus, King of Thrace.

the value of the metal, although such marks did not necessarily make the ingots more freely exchangeable, since they probably meant no more than that the man who issued the ingot would accept it back at the same value for which he had offered it in the first place. However, with standard quality metal made possible by the touchstone, and forgery easily detected by the same stone, the mark of the king's mint was now evidence of purity, weight and acceptability.

The need for smaller units of exchange took matters a step forward with the production of the Western world's first coin, the Lydian *stater.* Within a hundred years a set of coins, each one a fraction of the *stater,* had been issued. When Croesus of Lydia introduced the first standard imperial coinage in 550 B.C. Lydian money was already known for its high and unchanging standard. Other cities and states followed: Miletus, Phocaea, Cyzicus, Mitylene and Ephesus each founded official mints, and gradually their money began to be used and accepted outside the bounds of their own market-places, as monetary unions, like that between Lydia and Mitylene, were established. By the time of the Athenian Empire, in the fifth century B.C., money from Athens was accepted in most parts of the eastern Mediterranean.

As the use of coinage spread, it had two fundamental effects. The first was political: money issued by a central mint had a unifying effect on the users. The mark of the government on the coin was present in every transaction. Its presence defined the boundaries of governmental authority, and its value mirrored the health of the economy and the political stability of the country. The second effect of coinage was a consequence of the first. As the states developed and prospered, trade between them increased, and the use of coinage permitted more selective buying and selling of more diverse cargoes. Markets became more varied, and more widely scattered. It could be said that the introduction of the Lydian *stater* triggered the growth of trade in the Mediterranean because coinage made possible much more flexible trading methods.

From earliest times the world's great trading route had run from the eastern Mediterranean down the Red Sea, and across towards India and China. There is evidence of the Egyptians and Babylonians going as far south as Somalia in the second millennium B.C., and east towards India. Throughout the life of Alexander the Great this trade route was developed and held open as much as anything by his use of coinage. Alexander's mark was accepted from India to the Lebanon, from south Russia to the upper reaches of the Nile. In 331 B.C. he decided to found a city at the most convenient point to handle the flood of commodities criss-crossing his empire. The city was to be built in stone at a place where two natural harbours, facing east and west, would permit land-fall whichever way the wind was blowing at the time.

17

Left: Nobody knows what ancient Alexandra looked like, since little archaeological evidence remains. This drawing (c. 1460) is a medieval artist's impression of the city. Note the Pharos lighthouse built on the tongue of land jutting out at bottom right, in antiquity an island. The Pharos was so impressive that its name passed into Mediterranean languages as the word for lighthouse.

Alexandria, at the mouth of the Nile, became the greatest trading capital of the world. From the south, from Somalia, came spices. From the Sudan came elephants, iron and gold. From France, Germany and Russia came furs and amber; from England, tin. Alexandria took goods from all over the world and redirected them to their destinations. Dio Chrysostom said of it, in the first century A.D., 'The city has a monopoly of the shipping of the entire Mediterranean . . . situated as it is at the crossroads of the whole world.' After Alexander's death the city was ruled in turn by Persians, Greeks, Carthaginians and finally Romans. For six hundred years Alexandria prospered, both as a trading community, and as the intellectual centre of the Mediterranean – thanks to the great Library and Museion, founded not long after the city was built. Here, the greatest teachers of the time gathered to write and to give lectures, supported by public funds, in one of the ten halls, each devoted to one of the subjects taught on the curriculum. There were rooms for research and study, and quarters for teachers in residence.

The Library was a treasure-house of virtually all that was then known. In 235 B.C. there were almost half a million manuscripts there, and by the time of Julius Caesar the number had risen to 700,000. The collection was augmented through the implementation of a law which required all visitors arriving in the city to lend any manuscript in their possession to the Library for copying. Officials searched ships for books as they arrived. During the fourth century a quarrel developed with Athens when it was discovered that they were

Below: Ptolemy's view of the universe, consisting of concentric crystal spheres each carrying one of the seven known planets (including the sun and moon), and the outermost, the stars. In this scheme the Earth is placed at the centre of everything.

borrowing originals of the Greek tragedians, and returning the copies made by the Library staff. The subjects taught at the Museion encompassed every field of contemporary learning, including mathematics, geometry, astronomy, philosophy, medicine, astrology, theology and geography. Not surprisingly, since Alexandria was a seaport, special support was given to war studies, and to geography and associated fields of study such as astronomy. At some point between A.D. 127 and 151 one of the greatest scholars who ever taught in Alexandria, Claudius Ptolemy, wrote a thirteen-part work called *Mathematike Syntaxis* (The System of Mathematics), which brought together everything that was known at the time about astronomy.

Since early in the second millennium astronomy had been a field for study, under the Babylonians and to a lesser extent the Egyptians. At the beginning interest in the behaviour of the stars and the sun and moon had been purely practical, since it was believed that a calendar based on observations of the sky would record the seasons with greater accuracy, and this in turn would enable administrators to foretell times of flood for irrigation. Gradually the discipline took on the myth and magic of the astrologers, and predicting eclipses or the behaviour of disappearing constellations at different times of the year gave power to the priest-kings. By the time the Persian King Cyrus was called into the Babylonian heartland in 539 B.C. to save the country from civil war, astronomers had divided the sky into twelve constellations, 30 degrees apart, in a circle of 360 degrees, and laid the basis of the zodiac. The Persians, pragmatic realists, turned much of the Babylonian mumbo-jumbo to more scientific observation, and from 300 B.C. the Chaldean Tables were developed.

A mosaic from Ostia, the port of Rome, showing its lighthouse, perhaps the best known in antiquity after the Pharos. Although smaller, it was of similar construction, built in several storeys diminishing in size towards the top, where the fire was housed.

Prolemeus

20

These were the bases for Ptolemy's work. The accuracy they required in measuring degrees of position and minutes of time demanded the use of astronomical instruments. Ptolemy himself produced a Star Table, as part of the System, listing 1022 stars and giving their positions in the sky as seen from Alexandria. He also described the construction of an instrument for measuring these positions: the astrolabe, which was to become the basic tool of observers for the next thousand years. Though these star tables of Ptolemy's could have been of considerable use to the navigators of his mercantile city, there is little evidence that more than a very few astronomer-sailors ever took advantage of astronomical data either before or during Ptolemy's lifetime. Those who did, such as Eudoxus of Cnidus (370 B.C.) or Pytheas of Marseilles (300 B.C.), used the measurement of the angle at which a star lay in the sky to tell them how far north or south of their home port they were. Thus if a table said the angle should be x, and in fact it was y, the difference would relate to the position from which they were observing the star, and by working backwards from the table position they could work out where they were. However, for the most part Greek and Roman sailors used the stars to determine direction, and little more. As the Roman writer Lucan said in 63 B.C., 'We follow the never-setting Axis, that guides the ships. When the Lesser Bear rises and stands high above me in the yards, we are looking towards the Bosphorus' (in other words, with the Bear – to the north – on the left, the ship was heading east).

Three hundred years after Ptolemy wrote, Rome fell, and what little of the work of Alexandria that had penetrated the lands north of the Mediterranean was lost in the confusion and anarchy of Europe when the Romans withdrew. Western Europe as a whole lost contact with the Empire in the east, and although teaching at Alexandria went on under the Byzantine emperors, the importance of the Library declined. When the burning of the books occurred, no one knows. There are various possibilities: accidental fires during the period of Roman domination, destruction at the hands of fanatical Christian mobs in the fourth century A.D., or in A.D. 646 when the Muslims finally took Alexandria and, according to Arab writers, burned all the books in the hearths of the city. Whoever destroyed the Library, almost all the texts were lost. We know of Ptolemy's system only because of the strange journey taken by one copy, which by the middle of the eighth century had reached the library of a monastery of south-western Iran at a place called Jundi Shapur. In A.D. 765, shortly after the founding of Baghdad a few hundred miles away, the Arabs made contact with the monastery. They discovered the library, with its collection of manuscripts containing vast amounts of Greek scientific and philosophical material, and began almost at once to

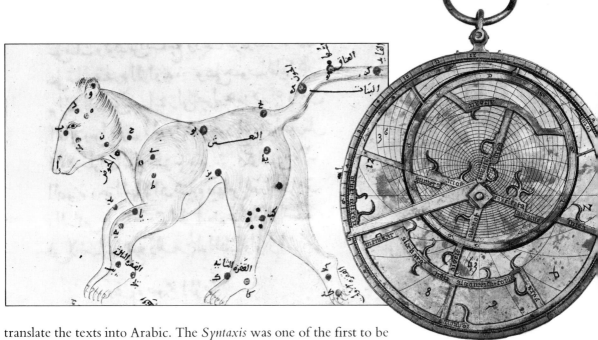

translate the texts into Arabic. The *Syntaxis* was one of the first to be translated. As the school of translation in Baghdad grew, the Caliphs bought or took from the Byzantines more and more Greek texts, especially those dealing with medicine and astronomy, and as Islam spread westwards into Spain, Arab translators and writers such as Averroes, Albumazen and Al-Kwarizmi brought the ancient texts and their own commentaries on them to the doorstep of Western Europe. So it was that Alfonso the Wise, one of the early Christian kings of what had long been Muslim Spain, was able, in the mid-thirteenth century, to set up a school in Toledo to translate the Arab texts into Latin. It was during the course of this translation that the star tables finally came to the West, and into the hands of navigators who would find a use for them.

In the thirteenth century European sea trade was just beginning to recover from the centuries of invasion and confusion that had followed the Western Roman collapse eight hundred years earlier. At the beginning of the eleventh century we hear of the first quays being built again since the Roman departure. Contact was being resumed between the Mediterranean and the North Sea as the economy of Europe began to recover and grow. Sea trade had lagged behind the general development. Master mariners still sailed, their ships as the Romans had, between May and September. From November to March the skies were uncertain, and no merchant would risk capital on a cargo which might easily get lost and founder in foul weather. During the sailing season, navigation at sea was still aided by Roman lighthouses round the Mediterranean, or by sailing according to landmarks. Closer inland, centuries of experience in sailing particular stretches of coast had given masters the ability to steer by soundings.

The constellation of the Great Bear, shown in one of the earliest Islamic manuals of astronomy (A.D. 1009), based on Ptolemy's Mathematike Syntaxis. Next to it is a brass astrolabe, the instrument developed by the Arab astronomers. Each astrolabe was designed to be used at a particular latitude. By rotating the metal network mounted on the backplate so that the points of the tiny spurs came to rest on the positions of the stars (which each one represented) as the observer saw them in the sky, dates and times could be read off the star tables.

Ships of the twelfth century would inch their way towards harbour aided first by the length of line it took to reach the sea bed, and then by the quality and composition of the sea bed itself that came up on the end of a weighted line with a bit of sticky tallow attached to it, and finally by knowledge of current and temperature and wind.

Sailing was also hampered by the design of the ships themselves. Just as no backer would finance a winter voyage, the same caution applied to radical changes in ship design. Too much was invested in a hull to risk anything that was not tried and tested. So it was that the old Roman square sail still dominated the seas off Europe for several hundred years after it need have done. Its limitations were simple, but far-reaching. With such a sail a ship could only run before the wind. If you wanted to sail south, and the wind blew north, you sailed north. This fact forced masters to wait for a favourable wind before setting out, a delay that cut down even further the number of ships at sea in an already short sailing season. While the square sail served the purposes of the Egyptians and the Romans in crossing the Indian Ocean, where the winds blow steadily one way for six months of the year and then switch direction for the other six, it lacked the manoeuvrability essential in the contrary winds of Europe.

A twelfth-century miniature shows the pilot, on the right, using a long sounding rod to navigate along a river in the English Fens.

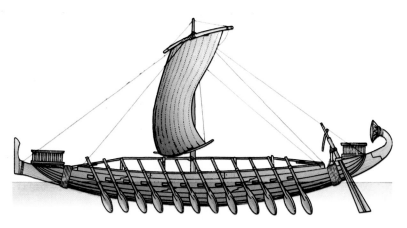

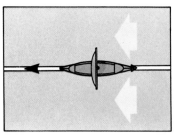

The device that changed things had been around the Mediterranean since perhaps the eighth century. It may have been Arab in origin, or the Arabs may have taken it from the Chinese. It takes its name – lateen – from the European misconception that it came with the East Romans from Constantinople. It was a triangular sail, fitted to a mast and a movable boom, and it permitted sailing close to the wind for the first time. Once under way, a lateen would catch the wind even when the ship was heading almost directly into it. The immediate effect of the spread of the lateen was to increase the number of voyages, since masters no longer had to wait for a favourable offshore wind before leaving port. The pace of trade quickened, and in consequence the size of ships increased, for as more and more cargo left harbour for more and more ports, it made sense to make one ship do the work of two. It saved money, and increased profit. As the northern European grain fields increased their output during the twelfth and thirteenth

Left: The typical square sail, used from Egyptian times. The diagram (above) shows how this type of sail left the ship's pilot totally dependent on wind direction, so that apart from their value in manoeuvring at close quarters in battle, oars were vital whenever it was necessary to make headway in contrary winds.

Below: Once this lateen-rigged ship was under way, the sail would provide enough thrust to keep the ship moving forward even when it was heading almost into the wind. A course set directly into the wind could be maintained by frequent coming about and moving forward in a series of zigzags, as the diagram below shows. The ship was still small enough to be manoeuvred by the use of a steering oar over the stern.

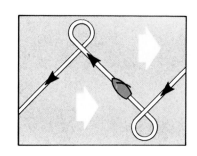

Above: One of the earliest representations of the sternpost rudder, on the city seal of the Baltic port of Ebling (1242).

Below: The first ocean-going rig: a mixture of square sail and lateen on a cargo vessel, easily manoeuvred with the help of the sternpost rudder. As the diagram shows, the use of the lateen sail permitted tacking through the varying offshore winds until the square sail could be hoisted and the ship could run before the steady transoceanic trade winds.

centuries, export of bulk cargo became commonplace. The northern states – and in particular the German Baltic Sea union of mercantile towns, known as the Hanseatic League – set up trading agreements with the southern, Mediterranean states for the exchange of commodities. Between 1150 and 1300 there was a growth of 30 per cent in prices, profits, accumulated wealth and capital invested in commerce. But there were still many problems facing the ships' masters, because even with a lateen rig the new ships built to carry grain were extremely difficult to handle, since they were the wrong shape to be run up on a beach or manoeuvred alongside a quay in anything but ideal weather conditions. That this was a particular problem in the north may be the reason why the first evidence of its solution is found on two mid-thirteenth-century city seals. Both of the cities were Baltic seaports: Ebling and Wismar. On both seals there is a clear picture of a ship with a rudder.

The standard method of steering had always been by stern oar, but, as the German masters found, there comes a point when the size of the ship makes the use of steering oars too cumbersome. The new rudder probably came from China, an idea adopted by the Arabs during their Indian Ocean trading expeditions, and gradually diffused to the Mediterranean and the north. Its full name is the sternpost rudder, since it projects from a post attached vertically in the centre of the stern, which must be flat to allow for the rudder attachment. So the new ships had a new look: pointed prow, square stern – a shape that, incidentally, provided more cargo space. The rudder gave the masters the necessary longitudinal control of their big ships, which in turn encouraged the merchants. More investment in cargoes followed. The ships became fatter, squarer, longer – and fuller.

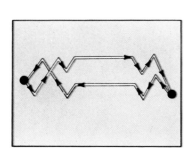

By 1300 a further complication had been added. Warriors returning from the Crusades brought back luxury goods such as spices, silks and dyestuffs that could not be bought in Europe, and in so doing created a new demand on the European markets. This was the driving force behind the early Italian maritime republics such as Venice, Genoa and Amalfi, which set up colonies and trading posts in Arab and Turkish lands in order to profit from the handling of the new luxuries. Many of the goods moving from the Italian ports to the north and west were paid for in kind: salt from the French Atlantic coast deposits, timber from Scandinavia, grain from Germany and, principally, wool from England and Flanders. But the European states began to suffer an imbalance of trade large enough for them to be concerned about the flow of their gold south. The problem lay with the middlemen: the Genoese and Venetians in particular, who were setting the prices and creaming the profits. Matters became substantially worse in 1453, with the fall of Constantinople to the Turks and the consequent loss of easy contact with Eastern markets. Prices rose higher still, and it was in reaction to this, and in the desire to find and market spices for themselves, that the Portuguese – encouraged by Henry the Navigator and his great map-making school of navigation at Sagres on the Atlantic – set off in the mid-fourteenth century down the coast of Africa in search of supplies. With those first African expeditions, the serious problems of oceanic navigation began to appear.

This late fifteenth-century illustration from Nuremberg of the construction of Noah's ark shows the shipbuilding techniques used in the period of maritime expansion in Western Europe following the fall of Constantinople.

Medieval navigators had relied on three aids, the portulan chart, the knotted line and the traverse board, to tell them where they were. Portulan charts (from the Italian *portolano*, a book of sailing directions) had begun to appear early in the thirteenth century. They were highly accurate maps of coastline – one of the earliest is of the Black Sea – carrying on them windrose lines. These lines represented the direction of the major winds of the area and were drawn to a scale, so that the navigator could measure with some accuracy the distance between two of the intersecting wind lines. The traverse board was used to plot the direction of the ship's course, by inserting a peg into one of a series of holes placed in concentric circles, with the ship's position at the centre. If the ship moved north-west a certain distance, the thread attached to the peg inserted at that point, connected to the peg at the centre, would give a visual representation of the ship's course at a glance. To determine how far and how fast the ship was moving, the knotted line was used. A line with a log on the end of it was thrown overboard. At regular intervals on the line was a knot, and as the log drifted astern the line was paid out for a specific interval of time measured by a sandglass (rather like a modern egg-timer). The number of knots paid out in that period would tell the captain his speed as a multiple of knots in an hour. With these three aids the master could work out his position by deduced reckoning (from which we get the modern 'dead' reckoning, a phonetic abbreviation). This technique, though rough and ready, sufficed for the length of journeys in the Mediterranean and along the coast of northern Europe. Once the transoceanic voyages began, however, some more accurate way of computing exact direction had to be found.

What made the African expeditions possible was a device that may have come originally from China, brought to Europe by the ubiquitous Arabs. There are early and sporadic references to it, the first by a travelling English monk called Alexander Neckham, who returned from Paris in 1180 with news of a mysterious needle that always pointed

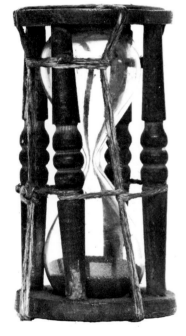

Two instruments for measuring a ship's speed. Above, a sandglass (used when the ship was moving fast and the knotted line was paid out very quickly) and, right, a log carved in the shape of a fish. The hinged portion brought the log to a halt in the water, so that the length of rope paid out from the moving ship could be accurately measured.

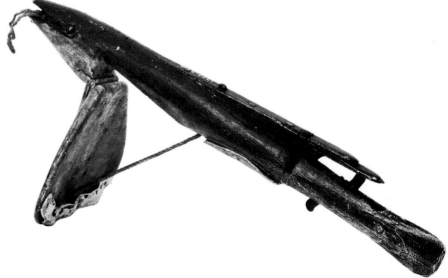

the same way. In 1269 another peripatetic divine, Peter of Maricourt, wrote of experiments he had conducted when he was with the Duke of Anjou at the siege of Lucera, in Italy, with something he called a 'dry pointer'. After 1270 Alfonso the Wise (of Toledan Star Table fame) had decreed that all sailors should carry the needle. By the end of the thirteenth century the needle seems to have been in general use throughout the Mediterranean. It appears commonly to have been stuck in a straw and floated in a bowl of water, but tradition has it that in 1300, in the Italian republic of Amalfi, the compass card had been added, showing all the points of the compass (derived from the directions of the major and minor winds). A frame had been added to hold the device in: the compass was boxed. The navigator could now plot his course accurately to within one-thirty-second of a circle, the wind directions having been subdivided into eight major winds, eight 'half-winds' and sixteen 'quarter-winds'. This standardization of the wind directions was born of Mediterranean experience, where the winds blow constantly and reliably in the same way.

The economic effects of the arrival of the compass can hardly be exaggerated. Together with the use of the combination of lateen and square sail, and the sternpost rudder, the compass altered the sailing schedule almost immediately. At the beginning of the thirteenth century the seas had been closed in winter, and sailing at that time of year was actually forbidden by law in the Italian maritime cities. From Venice, for example, the fleets left for the eastern Mediterranean twice a year. One left at Easter and returned in September, and the other left in August, to winter over in the foreign port and return in spring. A hundred years later the fleets still left as before, but the August departure was followed by a return through winter seas. With the aid of the compass, ships could sail under cloudy skies by day and night. The number of voyages doubled, and crews were kept in regular employment. This in turn encouraged the investors, and the number of voyages further increased. So it was that the fall of Constantinople, by cutting trade with the East, triggered the great trans-oceanic voyages that were to lead to the discovery of America by ships that could use their lateen sails to tack their way through contrary winds out to the Azores; there they picked up the steady trade winds in their square sails until once more the lateens would handle the winds in the Caribbean, do the same thing in reverse coming home, relying on the needle in all weathers to tell them where to go.

Tradition has it that the fact that the 'constant' needle was untrustworthy was originally noticed by Columbus on his first voyage across the Atlantic. In September 1492, when he was nearing the Caribbean, panic spread through his crew when it was observed that the needle no longer pointed at the Pole Star as they thought it should.

Above: The traverse board, used by sailors up to the end of the nineteenth century. The rows of holes along the bottom of the board were used to show how long the ship had travelled during each change of direction, each hole representing a predetermined period of time measured by a sandglass. The peg was inserted at the relevant point, row after row, so that one could tell at a glance how long each leg of the journey had taken.

Below: The boxed compass, showing the division of the circle into thirty-two parts.

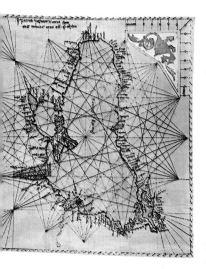

When he and others returned from voyages at the beginning of the sixteenth century with tales of this variation, there was immediate concern. If the needle could not be trusted, investment in voyages would no longer be safe, and the growing maritime empires of England, Spain and Portugal relied increasingly for their economic strength on the ability to navigate the seas with accuracy.

It is not surprising that the first experiments to find out why the needle erred should have been carried out in England, now the most dynamic of the new transoceanic European powers. In 1581 a compass-maker in London, Robert Norman, carried out a series of tests and published his findings in a book called *The New Attractive*. He had floated needles, both magnetized and unmagnetized, in bowls of water, and noted that, contrary to his expectations, the magnetized needles showed no inclination to float towards the north. He also noted that they *dipped* at their northern end. What no one knew at the time

Above: This reconstructed portulan chart shows how medieval navigators could calculate direction by the windrose lines, representing the local winds.

Right: The innovation of the lateen rig (at the stern) and sternpost rudder allowed ships like this fifteenth-century German carvel to make transoceanic voyages in all weathers. In just such a ship Columbus crossed to the New World.

was that there was a difference between true north and magnetic north, and that only when a mariner was in a position where the two lined up would his needle show true north. At all other times it would appear to point away from true north to a greater or lesser extent, depending on where he was on the surface of the Earth. Norman's revelations spurred investigation by a well-to-do doctor called William Gilbert, President of the Royal College of Physicians, and personal physician to Queen Elizabeth.

A lodestone mounted in copper.

Gilbert had taken his M.D. at Cambridge in 1569 at the age of twenty-five, and in the following year had moved to London, where he quickly became physician to many of the leading families of the capital. When he came to Court, in 1600, England had seen two long, stable reigns – that of Henry VIII and his daughter, the Virgin Queen. The country had won great victories against her enemies and, thanks to the judicious use of an effective secret police, had enjoyed decades of peace at home. Following the dissolution of the monasteries by Henry and the growth of overseas trade, England experienced a period of prosperity during which any man who was prepared to work hard with an eye to the main chance could become almost anything he wished, provided he kept his political nose clean. It was a community that we should now call 'socially upwardly mobile'. Life, for those who had some money to invest wisely in the expanding economy, could be spent in leisured pursuit of interests. For most of the upper middle class and aristocracy this meant weeks of hunting. For Gilbert, it meant trying to solve the riddle of the inconstant compass needle, and he took eighteen years to do it.

As a physician, Gilbert was the nearest thing to what we should call a scientist. He was also a personal friend of Sir Francis Drake, and most of the great voyages of discovery were made during the early years of his life. He corresponded with the scientific 'radicals' of the time both in England and on the Continent. From his studies at Cambridge he would have known of the classical authors' theories regarding the special curative powers of the lodestone, the rock that made needles magnetic, though he regarded most of these theories as nonsense. So, for eighteen years, Gilbert experimented with the mysterious lodestone. He carved tiny planets – he called them 'terrellae' – out of the stone, and brought them into contact with various materials, including metals, wood, water, amber, and magnetized needles. When he had finished, he published a book called *De Magnete* (On Magnets). When the book came out in 1600 it was an instant success, and within a few years it was being read all over Europe. Besides giving his own experiments, Gilbert brought together all that was known about magnetism. His major conclusion was that the Earth was a great magnet, with both north and south poles, spinning on its

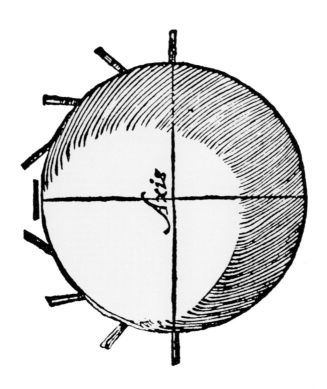

axis and travelling through space round the sun. It was the presence of a magnetic field, Gilbert noted, that prevented everything on Earth, including the atmosphere, from flying off into space. He also restated the fact that certain substances would become magnetic if they were rubbed. And, almost in passing, he said that if his conclusions about the atmosphere being held in place by the magnetic field were true, then there would have to be an absence of air between the planets.

These statements caused a sensation, and investigations were begun all over Europe to test his theories. The idea that excited most speculation was the possibility that a vacuum might exist, for in effect this was what Gilbert meant by an 'absence of air' between the planets. But according to Aristotle such a thing was not possible. 'Nature abhors a vacuum', he had said, because he thought that an object thrown or propelled through the air moved as the result of an 'impetus'. He believed that the air passed from in front of the object to behind it, thus pushing it. As the air became less dense, the passage of this 'pushed' object should become easier. If this were true, any object passing through a vacuum should encounter no resistance, and its speed would therefore become infinite. Aristotle could not accept the concept of infinite speed, and for this reason concluded that there could not be a condition that would permit it.

31

Much of the investigation into Gilbert's ideas was done by military engineers, who were interested in the relatively new phenomenon of gunpowder and what it would do to cannon balls shot from guns. One of these engineers was a German called Otto von Guericke, who was born two years after the publication of *De Magnete*. By the time he had finished his studies in mathematics and law at Leyden University in Holland (itself founded with a grant awarded to the town for surviving a siege), everybody was in on the act of trying to create the vacuum at will. By 1646, at the age of forty-two, Guericke was elected Mayor of Magdeburg, and had earned a reputation as a scientific dabbler. In 1652 the German Emperor Ferdinand III came to the southern German town of Regensburg to hold a meeting of the princes and mayors and bishops of Germany to discuss matters of policy. He had heard of Guericke's work, and commanded a demonstration, so in May of that year Guericke impressed the assembly with a vacuum pump he had developed. The pump was little more than a specially redesigned fire extinguisher, common at the time, with a system of valves which permitted it to evacuate various containers. One demonstration showed that a container with air in it weighed more than an evacuated container. Another showed that only spherical containers could be evacuated, since any other shape would implode. Within the evacuated container strange things happened: candles were extinguished, mice died, bells rang silently. But Guericke's *tour de force* was the horse experiment. He attached two teams of eight horses to either side of a pair of brass hemispheres, united and evacuated. Heave as the horses might, they could not pull the hemi-

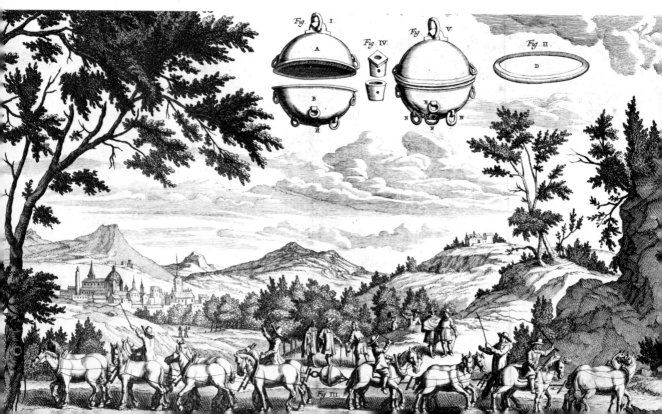

Above: A worker assesses the purity of sulphur by holding it to his ear and listening for the crackling produced by the warmth of his hand. Unless the sulphur is very pure no crackling occurs.

Left: To show the power of a vacuum Guericke invented a vacuum pump (above), based on fire extinguishers of the day, to empty the hollow 'Magdeburg' hemispheres (centre) of air. As a spectacular demonstration of this power, two teams of horses (below) were each harnessed to one half of a sphere: they were unable to force it apart.

Right: The sulphur ball conceived by Guericke was the first electrical generator. A hand held on the ball as it revolved produced frictional electricity which made the ball attractive to paper, feathers and other light substances.

spheres apart, until air was let into them, whereupon they fell apart at a touch. Guericke had shown that the vacuum could be created at will, and that its force could be made to do work. Ferdinand was impressed, and ordered the experiment to be written up by Caspar Schott, Professor of Mathematics at Würzburg University. Schott published two books on the experiments, the first of which appeared in 1657 and brought Guericke's efforts to the attention of scientific scholars all over Europe.

A few years later Guericke turned his hand to the other major area of Gilbert's work: magnetism. At some point before 1663 he began experimenting with the idea that certain substances would become attractive when rubbed. Sulphur was one, and Guericke built himself a sulphur globe by pouring crushed sulphur into a glass sphere and heating it until it melted. When it hardened again the glass was shattered, leaving the sulphur globe ready for work. Guericke mounted it on a rod, and set it horizontally in a frame, attaching the rod to a gearing system and a crank handle so that the globe could be spun at high speed. As it turned, he rubbed it with his hand and found that after a while it would indeed attract feathers, linen threads, water and so on. Guericke therefore concluded that it was this attractive force which pulled objects back to Earth after they had been thrown into the air. Then he did one more test. He rubbed the spinning sulphur globe in the dark and saw it glow, and watched the glow extend from the globe to his hand, placed a few inches away. To Guericke this light was just another strange aspect of magnetism, as was the crackle he heard when he rubbed the globe and placed it close to his ear. In 1672 he published the results of his work; in the book, *Experimenta Nova Magdeburgica* (The New Magdeburg Experiments), only one paragraph was dedicated to the sulphur globe. It was enough to set off a century of discovery.

In the history of the process of change there are certain crucial moments when the number of paths down which subsequent events can lead suddenly multiplies. Guericke's publication was one such moment. His work on the vacuum pump led to research into the composition of gas and in particular air. This led to the discovery of oxygen, which in turn led to work on combustion, respiratory diseases and the analysis of the elements. It also helped to solve the problems of mine drainage, and produced advances in metallurgy, notably steel production. The examination of gases would one day lead to the investigation of light passing through the gases, and that in turn to the discovery of cathode rays and the television set. His experiments with the sulphur ball did more. The force Guericke had seen at work, glowing and crackling, was electricity, and there is no need to detail the inventions and discoveries which resulted from that.

Perhaps the least obvious result of Guericke's work on the sulphur ball was to quicken interest in the weather. Men had for centuries speculated on the nature of thunder and lightning. In Saxon England the ecclesiastic Bede had speculated that lightning was due to the rubbing together of clouds, and thunder to the sound of the impact. In the Middle Ages it had been the custom to ring peals of bells in the church steeples to disperse the thunder, as a result of which a high number of bell-ringers had been electrocuted. Indeed as late as 1786 the Parlement of Paris enforced an edict forbidding the practice, because over the previous thirty-three years and 386 recorded lightning strikes no fewer than 103 unfortunates had been killed on the ends of their wet bell-ropes. Within thirty years of Guericke's experiments, the connection had been made between static electricity and lightning. In 1708 Dr Wall, in England, wrote that electricity 'seems in some degree to represent thunder and lightning', and in 1735 another Englishman, Stephen Gray, trying to find how far down a thread the mysterious force would go, came to the same conclusion.

Concern soon focused on lightning strike and the danger it presented to gunpowder arsenals all over Europe. The row over exactly how to protect them began with the work of a hitherto obscure fifteenth child of a Bostonian soap-boiler, Benjamin Franklin. In 1750 he wrote to the British Royal Society expounding his electrical theory, which stated that there were two kinds of electricity, positive and negative. The reason that electricity flowed from one place to the other – a phenomenon everyone had observed – was the desire for a negative material to move to a positive one in order to achieve natural balance. He claimed that electricity would therefore be attracted to a positive iron rod, and away from more dangerous, fragile or expensive properties that might otherwise lie in its path. He suggested that a church steeple be used to prove his theory. The Royal Society was not

Above: A bell-ringer of St Pol-de-Léon in Britanny is killed in 1718 by lightning.

Below: A lightning conductor for use in kite experiments, as illustrated in E. Chamber's Cyclopaedia, *London 1779.*

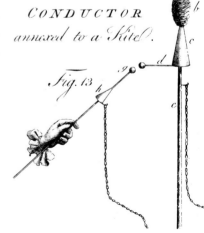

Below: Franklin's lightning conductors caught the imagination of Europe in the 1770s. In Paris fashionable ladies promenaded with conductors attached to their hats, and gentlemen availed themselves of the ultimate extravagance, a portable lightning conductor.

interested, so Franklin tried it for himself, and since the Philadelphia church spire he had in mind was not ready in time Franklin attached an iron wire to his now famous kite and was hit by the shock from a storm cloud. The following year, 1753, he published his data in *Poor Richard's Almanack*. In 1760 the first lightning rod was installed in England.

The explosion of an arsenal in Brescia, northern Italy, in 1769 made the rod a political issue. An estimated 175,000 pounds of powder exploded, destroying 190 houses within a radius of 639 feet from the explosion. The Brescia authorities asked the Royal Society for help in preventing a further disaster, and a committee was set up, of which Franklin was a member. An issue developed over whether the rods should be pointed at the top, as Franklin said, or round. The British settled for the round variety, on the grounds that Franklin was a revolutionary. Conductors sprang up all over Europe. There was even a *chapeau paratonnerre* – an anti-lightning hat – for the ladies of Paris in 1778.

Two events followed that were to put the investigation of the weather on to a scientific basis. The first was the 1793 war between Britain and France, when ships were forced to leave their posts at critical moments in the battles because of the risk of explosion during storms. The British Admiralty committee set up to solve the problem recommended copper strips on the masts connected to metal parts on the hulls, and further asked that ships begin to report on storm activity as soon as they reached port. The second event followed an earlier flight of a hot-air balloon built by two paper-makers (called Montgolfier) from central France in 1783. In Hamburg, in July 1803, a French scientist called Etienne Robertson and a music teacher named Lloest took off in an ex-Army balloon and reached the dizzying height of 23,000 feet, at which point the effects manifested themselves in a swelling of M. Lloest's head to the extent that he could not get his hat on. But they had proved that man could survive at the altitudes where storms grew, and there followed a plethora of aeronautical activity, taking up air collectors, barometers, thermometers, compasses, telescopes, and any other device that might prove useful to weather observation – including live animals. By 1823 the British Meteorological Society had been formed, and weather observation took on a public purpose. On 14 November 1854 a hurricane struck the pride of the French fleet anchored off the Crimea, and the ship, the

Left: In a famous experiment of 1752, Benjamin Franklin and his son drew electricity from storm clouds via a key attached to the wet, hempen string of their kite into a Leyden jar.

Right: A weather picture of the whole of Europe as it was recorded on the evening of 2 December 1861.

Below: The world's first airforce, the Aerostatic Corps of the Artillery Service, was formed in the French army on 29 March 1794. One of their balloons is seen here tethered over the battle of Fleurus, Belgium, on 26 June 1794.

Henri IV, went down with all hands. The French War Ministry asked the director of the French Observatory, a M. Le Verrier, to see if something could be done to provide some systematic warning of similar impending disasters. On 16 February the following year, Le Verrier told the Emperor that a network of weather stations was what was needed. The next day his plan received the royal assent. On 19 February he presented the Academy of Science with a weather map of France showing conditions as they were that morning at 10 a.m., made up of data collected from ten different locations throughout the country. The idea of publishing weather maps was immediately adopted, and in 1861 the first book of maps was produced by Francis Galton in Britain. They looked remarkably like their modern descendants.

EXPLANATION OF THE SYMBOLS USED IN THE WEATHER CHARTS.

RAIN.

CLOUD.

Rain.	Snow.	Entirely and heavily clouded.	Entirely clouded.	Mostly clouded.	Half clouded.	A few clouds.	Clear blue sky.

DIRECTION OF WIND.

∧ ∧ ⌐ ⊃ ⊃ &c.
S. S.S.W. S.W. W.S.W. W.

FORCE OF WIND.

Ѱ Ѵ Ѵ Ѵ ∪ O
Gale. Strong. Moderate. Gentle. Almost calm. Calm.

Towards the end of the nineteenth century weather stations were set up on the tops of mountains all over Europe and the eastern side of the United States for the purpose of gathering information for the new weather forecasts. The highest mountain in Britain is Ben Nevis, in Scotland, and there, on 18 October 1883, *The Times* reported:

> It was arranged that a procession of ladies and gentlemen should muster at the Alexandria Hotel [in Fort William, at the foot of Ben Nevis] and walk or ride the bridle path leading to the summit . . . Mrs Cameron Campbell mounted a pony and led the way, headed by a piper playing 'Lochiel's awa' to France' . . . long before the party reached the top they found themselves in an Arctic region – the snow being two feet deep on the summit, and strong wind sweeping round in fitful gusts, while the temperature was intensely cold. Here . . . the architect . . . received Mrs Campbell, gave her the key of the outer door of the observatory . . . and the party found a welcome within, with a comfortable fire and tea.

Lord Abinger, of the Scottish Meteorological Society, stated during the ceremony that with a high station in Britain to work in conjunction with those in America and France, regular forecasts could now be made on a wide basis. The Ben Nevis Observatory was declared open.

In July 1803 Professor Etienne Robertson and Monsieur Lloest soared in their balloon to the unthinkable height of 23,000 feet, where Monsieur Lloest's head swelled so much he could not get his hat on. Their achievement showed that weather observation was possible at the altitudes at which storms originated.

Above: The anemometer at the Royal Greenwich Observatory, as it looked in 1880. The Observatory was one of a series of weather stations across Europe, and the weather was considered so important that it became the responsibility of a special Meteorological and Magnetic Department.

Above right: The Ben Nevis Observatory had been opened as Britain's first high weather station in October 1883. It was here that in 1894 C. T. R. Wilson observed the glory that led him to the cloud chamber.

It was in this observatory, some years later, that an event was to occur that would lead to the development of a modern invention whose existence has the most profound effect on every living organism on the planet today. It happened because the observatory was short of funds, and recruited temporary observers from the universities during times when the regular staff were on holiday. So it was that a young physics graduate called C. T. R. Wilson came to the mountain in September 1894 on vacation from Cambridge, to work for two weeks. One day, just after 5 a.m. at the beginning of his session of hourly weather observations, he witnessed this phenomenon. As he later described it: 'The shadow of the Ben on the surface of the sea of cloud [below the summit] at first reached to the western horizon. Its upper edge came racing eastwards as the sun rose. On the cloud surface beyond the shadow would then appear a glory, the coloured rings incomplete and rather faint and diffuse. . . . this greatly excited my interest and made me wish to imitate them in the laboratory.' The glory Wilson had observed is caused by the behaviour of light when the observer's shadow is cast on to fog or cloud in a valley. The head of the shadow is seen surrounded by between two and five coloured rings, with the red end of the spectrum on the outer edge of the rings.

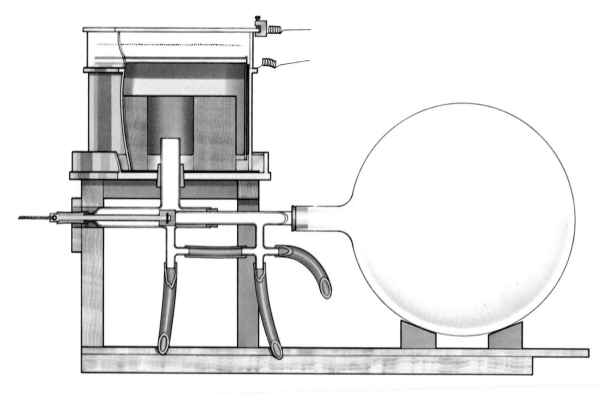

Wilson returned to Cambridge and began to try to make clouds artificially. In order to do so he designed himself a cloud chamber. The chamber worked on the principle that a drop in pressure would cause condensation of moist air. A glass container with a piston inside it was connected to a glass globe via a valve. The air in the glass globe was pumped out, and at the same time moist air was let into the glass container above the piston. When the valve between the glass globe containing the vacuum and the underpart of the glass container was opened, the air below the piston rushed into the glass globe, pulling the piston sharply downwards in less than a hundredth of a second. In the space above the piston, the sudden downward movement caused a drop in pressure of the moist air, which, as a result, condensed to form clouds. At the time, everybody thought that the condensation of water into clouds happened because the tiny droplets coagulated on to fine particles of dust in the air, and that this was what caused rain to form. But when one day Wilson used an electric charge to attract away all the dust particles which might have been present in the air within the container, and had done this for long enough to be sure that the air was absolutely free of dust, the impossible happened: condensation still took place. Months of testing showed that condensation had nothing to do with the presence of dust after all. Yet the faster the drop in pressure in the container, the denser the fog Wilson produced. At this juncture Wilson concluded that the condensation must be due to some particle too minute for him to detect, and there he let the matter

Above: C. T. R. Wilson designed his cloud chamber to recreate artificially the glory (see below) he had observed from Ben Nevis Observatory in 1894. Wilson developed the device by wiring up the glass container at the top to pass an electric charge through the cloud. He later passed X-rays instead of electricity through the chamber and concluded that the droplets were coagulating on ions.

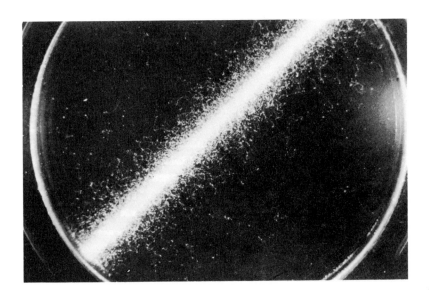

rest. He was not to know that in those tiny cloud streaks across his chamber he had triggered a scientific time bomb that was to explode with unimaginable force years later.

Meanwhile Wilson went back to his first love: the weather. In 1896 he used the newly discovered X-rays to produce clouds in his chamber, and decided that the droplets were coagulating on ions – atoms from which the negative electrons had been stripped by the passage of the radiation. Since this was what also happened in clouds, he concluded that lightning must be caused by the discharging of a giant difference in potential which built up between the top of the cloud, where the atoms were still positive, and the base where the rain droplets were negative. Between 1920 and 1925, while working out his theory of what caused lightning flashes, he was in contact with another Cambridge researcher, Edward Appleton. Appleton was interested in the measurements Wilson had taken to show the strength of the magnetic field in the immediate region of a lightning flash, because he was himself looking into the subject of radio atmospherics – the cause of crackle when a storm occurred during a radio transmission. He was also in contact with Robert Watson Watt, who since 1919 had been meteorologist-in-charge at the British Royal Aircraft Establishment, and who had been attempting to work out a way of using radio to locate storms. In December 1924 Appleton eventually discovered that radio emissions would bounce back from an ionized layer of atmosphere – now known as the Appleton layer – and that by timing the interval between emission and the return of the signal, the distance between the antenna and the layer could be measured. Watson Watt was able to refine Appleton's techniques for measuring the distance between a trans-mitter/receiver and the storm he was trying to locate, to the extent of

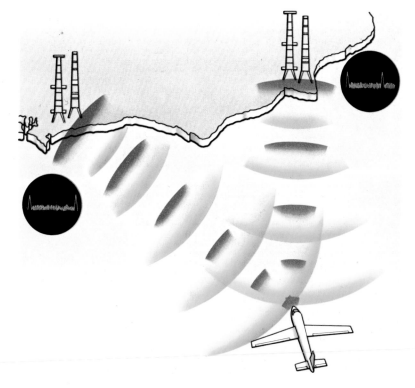

being able to give its direction and range. On 2 April 1935 Watson Watt received a patent for his Radio Detection and Ranging (RADAR) device which was capable of locating and ranging aircraft.

During the Second World War radar was used extensively to locate aircraft and storms. It was carried in a B-29 of the U.S. Air Force which took off on a mission on 6 August 1945; on board that day was the other product of Wilson's cloud chamber. Back in 1912 he had taken photographs of the streaks of condensing droplets in the cloud chamber, and had shown these to Ernest Rutherford, a colleague at Cambridge who was working in the field of atomic physics. The photographs showed the streaks and, at certain points, secondary streaks moving away from the main track at a tangent. Rutherford took one look at the photographs and became immensely excited. They showed what he had deduced less than a year before: the scattering of subatomic particles of alpha radiation.

Rutherford was to call these photographs 'the most original apparatus in the whole history of physics', for they provided the opportunity to study the behaviour of the atom under bombardment. It was this technique which aided physicists in their attempts to split the atom and which helped to produce the payload on board the B-29 Enola Gay on her mission in 1945, carrying the first atomic bomb to be dropped in war. The target was Hiroshima.

Three days after Hiroshima: the view over Nagasaki on 9 August 1945.

44

3

Distant Voices

When Enrico Fermi, an Italian immigrant to the United States, and his colleagues triggered the world's first atomic pile in Chicago in 1941, science opened Pandora's box. Out of it came new ways of healing, new tools with which to study the structure of the universe, the potential for virtually free electric power – and the atomic bomb. Of all the developments of atomic physics, two possibilities affect our future more than any others: electricity produced by the fusion process and annihilation by nuclear strike.

The fusion process makes electricity by producing immense amounts of heat with which to provide the steam power required to run electricity generators. It does so as a result of fusing atoms together, rather than splitting them as is the case in the fission process. This fusion can only take place at temperatures which are difficult to maintain. Once the operating temperature has been reached, however, the fusion of the atoms produces enough heat to cause the process to continue; a fusion reactor in continuous operation would produce temperatures close to those on the surface of the sun. This heat would instantly vaporize any conventional container, so the process must be conducted within a magnetic field, whose charge would repel the tremendously hot particles. (The shape of this field is that of a bottle, and so the container is known as the magnetic bottle.) As yet scientists have managed to reach the necessary temperature for only a limited

Two pages from a military roll compiled in England sometime before 1448. The purpose of these rolls was to identify and register the heraldic coats of arms of noble families. By this date the possession of heraldic blazons had become hereditary. Their original twelfth-century purpose had been merely to identify the rider, encased in armour, to protect him from accidental attack by his own men in the heat of battle.

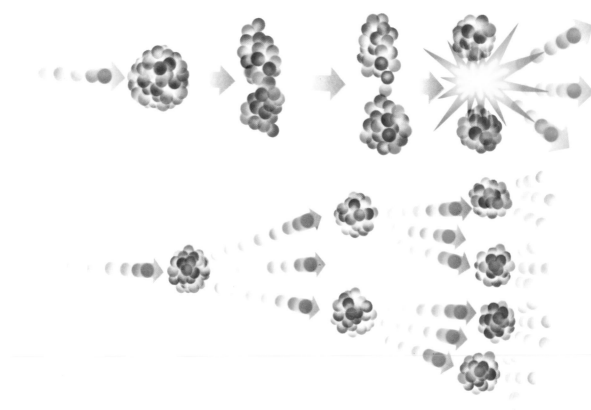

amount of time. If, as is expected, the process is in successful operation in the early twenty-first century, fusion will provide enough electricity to run a city the size of Los Angeles on the atoms contained in a bucket of seawater. The generation of electricity from such an abundant source of fuel will have the same effect as an enormous increase in the amount of other raw materials, since with unlimited energy it will be possible both to produce substitute materials which were previously too costly, and to exploit raw materials which hitherto did not justify the expense of their recovery. Energy is the ultimate currency, and fusion power would bring the opportunity of utilizing the vast untapped areas of our planet to feed, clothe and house mankind on a scale undreamed of in pre-atomic times.

None of this will happen if the second option, nuclear war, ever occurs. If it does, the massive destruction that ensues will also be caused by the fusion process. Until recently the only process which has operated successfully was the uncontrolled version – the hydrogen bomb explosion, in which the heat necessary to set it off is that produced by the detonation of an atomic bomb. So far, the possibility of such a device being used has been minimized by agreement among the few nations capable of producing it. But as the atomic era

Nuclear power is derived from the energy released when the nuclei of certain atoms are divided (fission) or fused (fusion). The atoms of very heavy electrons contain large numbers of protons and neutrons, and bombarding the atoms with neutrons makes atomic fission (above), so creating new atoms and releasing spare neutrons to continue the chain reaction. The energy produced is many millions of times greater than that of the rearrangement of electrons involved in chemical reactions. Uranium 235 (the isotope of uranium containing 92 protons and 143 neutrons) is the natural element able to

lengthens and the knowledge of how to build and operate atomic power stations spreads, the means to produce atomic weapons is becoming available to smaller, less politically stable countries. The secure monopoly once exercised by the major powers no longer exists. In particular, the complex and costly defence systems which up to now have acted as deterrents can today be breached with ease by any physics graduate with access to the necessary radioactive material. Anti-missile batteries are impotent against attack by a suitcase bomb, and, as the technology of miniaturization improves, there is no reason why a nuclear device should not be contained in something the size of a handbag. Such a weapon delivery system is virtually undetectable, and its use would radically alter the future.

Few breakthroughs in military technology have had similar potential for altering the society that first uses them. One such breakthrough occurred in Europe just over a thousand years ago. On that occasion the device was also small, both sides possessed it, one side used it first, and the results of its use were radical and far-reaching. Indeed, had the device not been used this book might well have been written in a different language.

The first effect of the device was to change the government of England. At dawn on a bitterly cold day – Friday 13 October 1066 – Harold, Saxon King of England, and his exhausted army finally pushed their way through the last few miles of dense forest that covered most of south-east England. They came out on to the open chalk hills that stretch to the English Channel and found themselves three miles from a place called Caldbec Hill, a rise overlooking the village of Hastings. Harold and his men had marched virtually straight from their battle with the Danes at Stamford Bridge in Lincolnshire, 270 miles to the north, in the quite extraordinary time of ten days. To make up for their losses at the battle they had gathered levies of men from the counties they passed through on their way south. Even so, when Harold pitched camp on the ridge above the sand lake (Senlac) that stretched along the foot of the hill inland from Hastings, he had only a remnant of the host of men he had taken against the Danes two weeks before. He may have had at most five thousand trained fighting men. Many of these would have carried the iron-tipped ash spear, or a throwing axe, and a sword; some were bowmen, most wore leather caps and very few had mail shirts. Ranged alongside these professionals were the levies: farmers and peasants, for the most part, who had been straggling in from all over the southern counties during the previous few days.

The ninth-century Saxon man-at-arms merely used his horse to ride to the scene of action, in this case to surround a fortress. His helmet was made of hardened leather but only protected the crown of his head for fighting on foot.

Against this motley group stood the troops of Duke William of Normandy, later to be known as the Conqueror. The Normans had landed sixteen days before, and after having laid waste to the coastal areas they had since had over two weeks in which to rest and prepare for battle with Harold. William had over eight thousand fully trained fighters: 1000 archers, 4000 men-at-arms and 3000 cavalry. All of them were both well fed and well trained. They used standard weapons – lances, spears, axes, bows and arrows, swords – and most of them wore steel caps and mail shirts.

During the night of the thirteenth William offered terms: he would rule the south, and Harold could retain everything north of the river Humber. After some discussion in the Saxon camp Harold refused the offer, and battle was inevitable. Although Harold was unaware of the fact, he had already lost. For in the ensuing battle the Normans were to use a device of crucial significance that had been perfected in northern France over the previous hundred years. Harold and his predecessors knew of its existence, and indeed used it themselves, but the conservative fighting style of the Saxons may have prevented them from realizing its full potential.

The twelfth-century fighting man used his horse to fight from. His helmet protected the whole of his head and he wore more sophisticated battle gear of a mail coat. The men-at-arms behind are similarly equipped.

Left: The Norman cavalryman (as he looked at the battle of Hastings) carries a kite-shaped shield on his left arm to protect the length of his body, and a lance in his right hand. His sword is for use after the charge in the mêlée. His ability to use the lance and sword while keeping his balance depends on his being able to stand up in the stirrups.

As the battle began, the English were grouped along the top of Senlac ridge, along a front stretching for 800 yards. The trained men stood at the front, looking across at the Normans on a slight rise to the south. As they watched, in the cold light of that October morning, they shuffled into position so that the shield wall stretched unbroken along the length of the front. The shield wall had been the standard defensive tactic used by the foot soldier against cavalry since late Roman times. Harold's men were on foot, because that is the way they had always fought. To ride a horse into battle would have been to come to the field already prepared for flight, and such was against the Saxon battle ethic, which decreed that if a man's lord died on the field, he could not leave it alive. So what use was a horse? The horses

were tethered up beyond the top of Caldbec Hill. Behind the front line came the levies, the farmers with their billhooks, hammers and pitchforks. A few had axes and daggers.

At 9 a.m. the Normans advanced, under a hail of arrows from their own bowmen. The plan was to break the English line with the arrow strike, and cut into it with the cavalry as the arrows took their toll. It failed. The Normans advanced up the hill towards the shield wall, and were beaten back with an avalanche of spears and axes. At 11.30 came a pause, during which the combatants caught their wind and collected some of the weapons they had thrown. At around 2.30 the battle began again. Once more the Normans tried to ride through the shield wall; they charged *en masse* up the hill at the centre of the English line and fell back, exhausted, after battering at it for more than two hours. As the Normans retreated into the valley, the English made their great mistake. They broke ranks and followed the Normans down to level ground. At that moment the Norman cavalry turned and, standing in their stirrups, cut the Saxons to pieces beneath their horses' feet. Here, where the ground was flat, the Norman cavalry shock-troop went through the English mass like a hot knife through butter. Twenty of the Normans rode at the English royal standard at the top of the ridge. In the desperate, hand-to-hand slaughter that followed, each of Harold's bodyguards went down before him in true Saxon tradition, taking sixteen of the Normans with them. The four survivors hacked Harold to death. By 5 p.m. on 14 October 1066, England was Norman.

The battle of Hastings was recorded eleven years later on the Bayeux Tapestry. To modern eyes it is a crude, childish work. It reveals the clues to the device that William knew how to use and Harold did not – and that ultimately gave England to the Normans. The device itself is hard to see on the tapestry, but its presence is apparent from something else that can only be there because of it: the kite-shaped shield carried by a rider whose right arm is occupied holding a lance, and who is therefore too busy to protect his vulnerable left leg. The fact that the shield is long enough to protect the entire length of the body reveals the extent to which the right arm is busy, and the only thing that would keep it so busy is a lance. And the lance is there only because of the device in question: the stirrup. It was William's use of the stirrup to build a shock-troop of cavalry that gave him the ability to ride down the Saxons once they were on level ground.

Nobody knows exactly where the stirrup first came from. The earliest illustrations come from several places in northern India – Sanchi, Pathaora, Bahaja and Mathura – where, in the second century B.C., there appears to have been some kind of loop into which the rider's big toe fitted. The wave of Buddhist missionaries carrying

Below: The Bayeux Tapestry is a blow-by-blow account of the battle of Hastings woven in 1077 by William the Conqueror's sister. The Norman cavalry charged down the Saxon infantry, some of the riders (as here) using the lance in the new horizontal position.

Indic culture northwards may have been responsible for the stirrup's next appearance, in the province of Hunan, China, in A.D. 523, when it was designed to house the entire foot. By the time the Arabs arrived in Persia in the late seventh century, use of the stirrup had spread as far west as Turkestan, and from there it followed an as yet unknown route into Western Europe, where it appeared in the kingdom of the Franks during the eighth century. The evidence for its arrival is only indirect, but it provides an interesting example of one of the peripheral effects of invention, that of language change. The Frankish words for getting on and off a horse were replaced at this period by words that mean 'to step up' or 'to step down'. Other effects show themselves: the *francisca*, the battle axe, and the *ango*, the javelin – both weapons predominantly used by foot soldiers – disappeared from the Frankish armoury. The *spatha*, the sword, became too long to be used with ease by a man on foot. And, most significantly, the spear took on a wing attachment just behind the head. This modification would only have been necessary if the spear was to become embedded too far in the body of the enemy to be easily pulled out. The force being applied to create this problem in the first place could only come from the impact of a horse. The spear was obviously being used as a lance, and this would only be possible with the stirrup acting as an anchor to keep the rider on the horse at the moment of impact. William's shock-troop was still in the experimental stage, since in the Bayeux Tapestry some of his cavalry can be seen using the spear–lance in the old way, standing in the stirrups and striking down with it. The value of the shock-troop was quite clear, however, and further experimentation soon followed.

The next stage of development came some time between 1040 and 1120. The cantle at the back of the saddle was raised like the back of a chair to give the rider further support; this encouraged even greater

Right: In the thick of battle there was little room to manoeuvre, whether one was mounted or on foot. The crush of fighters made it imperative that the knights were boldly distinguished from each other. Here at the violent battle of Roncevaux Roland is shown splitting the head of Marsile (extreme right).

Right below: The ornate armour, helm, crest and heraldic trappings of the fifteenth-century knight. His sign of identity in battle was now established by heredity as a family coat of arms, or crest. This is Thomas Montague, Earl of Salisbury.

Left: Watched by the ladies at their castle windows an early fourteenth-century knight is lanced and unhorsed, despite the heavy mail protecting his body.

speeds at the moment of strike, which in turn put stress on the single saddle girth. In the later twelfth century this problem was solved by the simple means of doubling the girth strap. With increased security of tenure, the rider was able more and more to concentrate on achieving maximum force of impact, and less and less on his own protection during the charge, which led to the development of armour to protect his body. In the early stages plates of steel were riveted on to his mail shirt, and by the end of the twelfth century he was almost fully encased in metal. Now the effect of the impact began to devolve on the horse. Up to then the animals were more in the nature of large ponies, and with several hundred pounds weight on their backs and an impact force of about 2000 lb, the maximum point of stress was transferred to the horses' front legs. Bigger horses were needed. By cross-breeding, an animal was produced that looked much like the modern shire horse, and the weight of these huge beasts added greatly to the force of impact. During the centuries of development the knight on horseback had become still further enclosed in metal, to the point where he needed some form of identification in battle to rally his men, and indeed in the thick of action to protect himself from his own side. At first this identification took the form of painted designs on the shield, for the most part geometric, using the nails in the frame of the shield as an outline. Later, as more and more men took to identifying themselves similarly, the designs became increasingly complex and decorative.

As each improvement in equipment was made, the business of being a mounted man-at-arms became more and more expensive. The rearing of horses for the early shock-troop formations demanded so much land that the kings are thought to have expropriated land from the church, and these 'ranches' were set up on the scale necessary to provide the number of horses and trained men required when the king went to war. Such a ranch would train a number of mounted men and be responsible for equipping them, under the command of the leader appointed by the king; this man would receive the necessary land in grant. If the parcelling out of land did in fact take place primarily for the raising of horses, these ninth-century ranches laid the foundation for later, feudal society. Be that as it may, by the fourteenth century the cost of equipping and maintaining a knight was considerable. Only those who could afford the land and the servants and the raw materials could become knights. The amount of money and possessions necessary to equip a mounted man-at-arms put the aristocratic rider literally and metaphorically above the rest of society. He was separated by his armour. He now also took on a new kind of identification: an unchanging surname. By 1300 the upper classes had dropped the patronymic, and had stopped calling themselves names such as John son of Ralph son of Geoffrey of Farnham. He

'How a man schall be armyd at his ese when he schal fighte on foote.'

would by then have called himself simply John (of) Farnham, and his family would have taken on the heraldic sign on his shield as a mark of dynasty. What had begun as a symbol of identity in battle had by now become the permanent family crest, useful in many ways in a world of illiterates. With the family crest, the social separation of the rich from the not-so-rich was now publicly and visibly complete.

By the middle of the fourteenth century the knights that these families were supporting in battle looked like tanks. Their impact – both physically and psychologically – on a foot soldier dressed only in a leather or mail shirt must have been considerable. The knight's head was encased in a helm, with a padded interior to take the shock of a blow. Forward vision was possible through slits in the helm, or through the use of a movable visor which was lowered in battle. Under the body armour the knight wore a close-fitting linen shirt, short breeches and long stockings. Over these he wore mail sections covering the crotch and upper leg; these sections were covered with plate armour, jointed at the knees and running down to the ankles. The feet were covered by jointed plate-armour shoes, called sabatons. The knight's body would be covered by a mail shirt, called a hauberk, with plate sleeves strapped on to it with leather belts, and the breast and back would be protected by a coat of plates. The outermost layer of clothing was a surcoat, often padded and embroidered with the family crest, and held closed by the swordbelt. His hands were protected by jointed plate gauntlets. Dressed like this, the knight weighed about 200 lb, which, when placed on the back of an armoured Percheron horse whose backbone was seven feet off the ground, was a formidable war machine when approaching at a gallop.

Below left: The knight of the twelfth century. The cantle of the saddle has been developed with a higher back and front for greater support at the moment of impact with the enemy. This extra stress and the added weight of the knight's metal armour has made necessary much stronger girth straps securing the saddle on the horse.

Below right: The knight of the fourteenth century. He and his horse are both by now massively armoured with mail and over this with metal plating. His horse has been specially bred for war and stands up to 21 hands (about 7 feet) high.

Small wonder, then, that the mere possession of knightly equipment was sufficient to alter a man's social position. Throughout Europe, many of the younger sons of noble families armed themselves by killing a man who was already equipped. In an attempt to prevent this medieval hooliganism tournaments were established, where the fatal exchange of equipment could happen under supervision. Tournaments had been introduced during the eleventh century, but were banned in 1130 by Papal edict on the grounds that too many well-trained knights were being killed. By 1316 Pope John XXII had reinstated them, if for no other reason than that they had continued for two hundred years despite the interdict. They were, after all, valuable training, they kept the aristocratic hotbloods off the streets, and as the years passed they became somewhat less bloody. By the fourteenth century it was sufficient to unseat a rider to claim his equipment. Some knights made this their full-time occupation, and one such – the fourth son of John fitz Gilbert, castelan of Marlborough – bettered himself to the extent of starting as an unknown man-at-arms and ending as Marshal of England. These tourneys have been portrayed in modern times (especially by Hollywood) as great events, attended by thousands

The French nobleman Jean de Saintré (left) jousts before the King and Queen of Spain and his mistress Madame des Belles Cousines, in fifteenth-century Barcelona. The heads of both combatants' lances have already splintered off in the first engagement.

56

A king and his army prepare to leave camp. Note, on the back of the mounted man-at-arms (top right), the leather straps securing his body armour.

from king to serf, rich with pennants fluttering in the breeze and gaily caparisoned horses prancing outside tents hung with bright shields, while within the knightly owner composed love poems and hung his lance with the lady's silken favour. The reality makes better reading. More often than not the occasion resembled a country fair, with more straw than silk, and more craftiness than chivalry. Many of the references to tournaments come in the form of complaints at the destruction of local property when the thing got out of hand, and the fighting spilled out into the streets and gardens of the towns. However, the tournament served its principal purpose of training a rider for his only aim in life: to hit a target at full gallop, so as to break the enemy line.

Going to war with men like these was a costly affair, in many cases bringing belligerent princes to bankruptcy. A knight would rarely embark for battle without the six horses which he was permitted, or his page apprentice to help him on with his armour. Providing enough food, shelter and material for a large number of these knights and their attendant foot soldiers was, by the fifteenth century, a far more complex and sophisticated job of administration than is generally imagined. The logistics were considerable, involving smiths, armourers, painters, tent-makers, fletchers (arrow-makers), cordeners (leather-workers), bowyers, turners, carpenters, masons, wheelwrights, saddlers, purveyors (of food), quartermasters and farriers. There were also surgeons, chaplains, legal and clerical staff, trumpeters and pipers, and, most important, cooks.

Osmodini vero idem·

Guillermus dux in die natalis domini ab Aexlando

This already complex business was rendered all the more so if the king or the prince himself came to the field. In the late summer of 1415, just such an army turned up outside a small village in the area which is now the Pas de Calais, in northern France. Once again, a battle was to be won by one side which had the technology to render virtually obsolete that used by the other. The large army that halted outside the village was French, and its purpose was to fight off – or end for ever – the claim by the English throne to the French crown. The man who had come to France to press that claim was Henry V. The battle between Henry and the French that followed, at Agincourt, was to prove the beginning of the end for the mounted, fully armed knight.

At dawn on 25 October both armies had been in position all night. The French had chosen the worst possible position from which to strike: they were between two woods, Tremecourt and Agincourt, which stood about three-quarters of a mile apart, but closed, at the point where the French would meet the English, to about half a mile. Into this gap would go all 25,000 French. As dawn rose, the French were in no state to fight. It had been raining all night, making the battlefield sodden with heavy, clinging mud, and most of the French knights had spent the night in the saddle in order to keep dry and were now standing around to keep their armour from getting muddy. The English were not much better off. In the previous seventeen days they had ridden or walked the 270 miles from where they had landed, with only one day's rest. For eight of those days they had been carrying heavy stakes cut from the woods to the south. It had rained most of the way and they had had little to eat but nuts and half-cooked meat. Many of them were suffering from bronchitis and dysentery. And they were outnumbered four to one by the French.

The battle of Crécy (1346) first showed the superiority of the Welsh longbow over the cross-bow. By the time the crossbow-man had cranked his bowstring back into the firing position and inserted his quarrel, the longbow had struck – as shown, in this contemporary illustration, by the dead French soldier, bottom left.

The armies took up their battle formations. The French, whose 25,000 included 15,000 mounted knights, drew their riders up in five ranks, the first two ranks dismounted, with a few crossbowmen in among them. The English formed three groups, four ranks deep. of dismounted men-at-arms, with wedges of archers between them. On the wings, facing inward, were two more groups of archers. For four hours nobody moved. The French knights were arguing about whether or not to charge, and by 11 a.m. there was a lot of jostling and pushing as the differences in rank and region began to show. No knight wanted to be in the second rank at the charge. Insults were exchanged, and arguments flared as the motley nature of the army, drawn from all over France, became clear. Meanwhile Henry had moved his men forward to within bowshot of the French, about 300 yards away. The stakes the English had carried for eight days were stuck in the mud, angled towards the French, points up. Still the French shoved and muttered, but did not move. Henry decided to make their minds up for them, and ordered his archers to fire into the air. Arrows from a thousand bows rained on the French, galling the

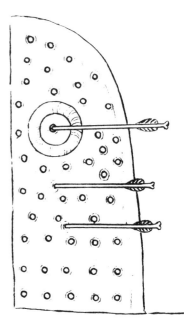

horses and wounding the tightly packed mass. Suddenly, the French charged, apparently without any central order, straight across the mud at the English. This time Henry's archers fired for the horses, bringing down riders in their hundreds. Many suffocated in the mud, unable to move in their armour, as their compatriots piled on top of them; many others were dispatched by English archers running forward to slide a knife between the joints in their armour. In half an hour it was over. The English had lost 500 men; the French 10,000. The myth of the invincible knight was shattered.

The weapon that had so suddenly turned the medieval social order upside down was the Welsh longbow. It had been introduced by Edward I, and was a formidable weapon that could shoot a rider at 400 yards, and with special steeled points would even penetrate armour at close range. An experienced archer could loose nine arrows a minute, and, as the grim jest of an English writer put it, when the French would turn to show their backsides to the English in disdain at the bow, 'the breech of such a varlet has been nailed to his back with an arrow, and another feathered in his bowels before he should have turned about to see who shot the first'. Fully three hundred years after Agincourt the longbow was still considered by many military experts to be the finest weapon any army could wish for, and yet within decades of the battle not enough archers could be found to fight a war.

Sunday archery practice at the butts in fourteenth-century England. Two targets were used, one at each end of the field, to save return journeys, and the target arrows were so blunt that there was little chance of anyone getting hurt.

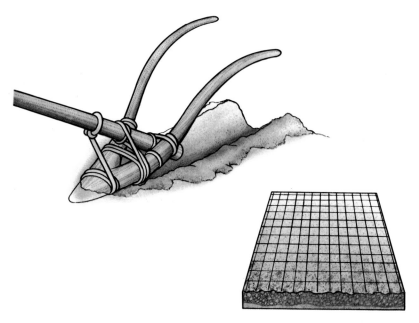

Left: The Mediterranean scratch plough, little more than a stick fitted with handles and a pole for yoking it to oxen. Used in lands where the soil was light and vegetation sparse, it could either be dragged to the field or carried by the farmer. The diagram shows the shape of field it worked.

Below: The forests begin to disappear as assarting cuts deeper into the woods. Note the vegetables in the garden and the men beating acorns out of the trees for the pigs to eat.

The reason the great bow failed so soon was that the face of Europe had been transformed in the centuries before Agincourt by three agricultural inventions, bringing with them a new prosperity for all. For thousands of years the plough in general use had been the Mediterranean scratch plough, so called because it consisted of a blade which simply split the soil as it passed. This plough was ideally suited to the land from which it had sprung, in the valley between the Euphrates and the Tigris rivers, in what is now Syria. For three or four thousand years before Christ it had broken the light, dry soil of the Mediterranean lands, pulled by oxen to criss-cross the field until every part was opened. The fields it ploughed were, in consequence, square. As Europe began to recover from the great plagues of the sixth and seventh centuries, and the population began once more to increase, the full effects of the plagues and the withdrawal of the imperial administration began to be seen. The forests had grown back, and the great estates had fallen totally into disrepair. In the tiny hamlets the problem was that the old scratch plough was not strong enough to carry out the work of clearing tree roots and tangled vegetation from the thick clay soil of northern Europe.

Then, in the sixth century, possibly from somewhere in central Europe, came a new plough that was to alter the face of the continent and the condition of its inhabitants. Nobody knows where it came from, though some historians place its origins in the Slavic countries because of early Slav words that may refer to it. Be that as it may, the new plough made its first appearance in western Europe in the Rhineland and in the Seine basin. The earliest clear illustration of it comes from a book of psalms written in England in the fourteenth century.

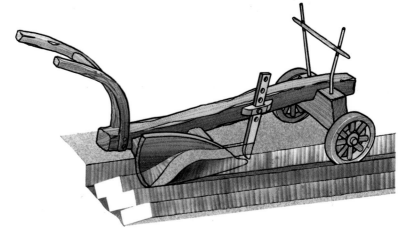

The medieval mouldboard plough. Apart from the plough's ability to open up the land in deep furrows, the iron knife could cut through the deep-rooted vegetation of northern Europe. The diagram shows the pattern of furrows and the new strip fields created by this plough.

The plough has wheels on the front, so that it could easily be taken from field to field. The major innovation lies in the way the plough cuts. On the frame ahead of the share is a knife, placed vertically to cut the sod and make it easier for the share to enter the thick soil. Behind the share is a curved, wing-like board, sitting diagonally to the frame, to lift the cut sod and throw it clear to the right, like a wave breaking. This was the mouldboard, and it may be said that this single fitting was the main cause of the agricultural revolution that was to follow, since it permitted land to be ploughed at its wettest. It did so because as each furrow was ploughed, the plough team would make a U-turn at the end of the field, and then return to make a furrow parallel with the first, going in the opposite direction. The result was a series of ridges and furrows, with the highest point of the ridges at the centre of the field, running like a kind of spine down the centre of the ploughed area. This allowed the water in the field to drain off to the sides.

It soon became clear that the new plough demanded more traction power, in the form of oxen, than any single peasant could afford – some of the ploughs took eight oxen at a time. The peasants began to pool their resources, and as it was more sensible to keep this pool of animals and equipment in a place central to the fields, they began to group together in villages. The new plough also changed the physical appearance of the land. Since it was so heavy, it made sense to cut long furrows and turn the teams less often. Thus the shape of fields changed from square to rectangular: the so-called strip fields. These were often cut into the forests, since doing so was easier than adapting the old square fields. Apart from anything else, the legal problems of adaptation would create problems that could be avoided by 'assarting', as the opening up of the forests was called. The chronicles of the great forest assarts read like the literature of the opening of the American West

a thousand years later. Whole village communities of serfs were moved across country to open up tracts of forest. One such group appears in the *Chronicle of the Slavs*, written in the seventh century. They were to be moved from a village in what is now Holland to clear land on the Baltic coast. As they set out through the almost impenetrable forest, prayers were said for their safety in the unexplored darkness between their homes and their destination. Most of them reached the Baltic. None returned.

Between the sixth and the ninth century, the new plough was joined by two more inventions. The first solved the problem of the oxen. They were slow animals, expensive to maintain, and it took a number of them to pull the plough. The horse would be much more efficient, but the problem was how to harness it. Ox harness, which was made up of straps across the belly and neck of the ox, would strangle a horse, pressing as it did on the jugular and windpipe. The horse collar, which was to put the horse in front of the plough some time in the eighth or ninth century, may have originated in Bactria in the sixth century as a form of harness for camels. Whether it came to Europe north from the Arabs, or south from the Vikings (some of whom had seen service in the Byzantine Empire and may have brought the idea back with them) is not known, but by the tenth or eleventh century the new harness was firmly established. It made an immediate difference, for the horse can work twice as much land as the ox. Production increased as a result, and therefore so did the population.

The other innovation that came to agriculture at much the same time as the harness was the nailed horseshoe, and its origins are equally doubtful. Some dated in the ninth century have been found in Siberia, and there are references from the same period to their appearance in Byzantium. By the eleventh century the horseshoe was in use in Europe. Its effect was to put the horse to work in all weathers and over all terrain, for it protected the animal's feet against both rough ground and foot rot. The horseshoe also permitted horses to be used more for transport, and because people could now ride their draught animals to the fields at a reasonable speed they could afford to live further away from them. The villages grew, as the peasants sought the benefits of community living.

It may have been the effect of the plough and the need to feed more draught animals as its use spread that gave rise to a practice that some think caused the centre of European wealth and power to move north of the Alps. The practice was that of running three fields at once. For centuries it had been customary to run two fields: one to grow crops on, and one left fallow for the animals to graze and fertilize with their droppings. With an increase both in population and in the number of animals being used, it may have been the need for more food than could be provided simply by assarting that caused the change to three-field rotation. It may also have been the fact that, unlike the Mediterranean, the north has enough rain in the summer to water an extra crop. Whatever the reason, the new system was to sow oats and legumes (beans or peas) in the spring, and cereals in the autumn, letting the third field lie fallow. Until then the population had lived mostly on cereals, and so legumes were an important addition to their diet. (Small wonder that one eminent historian has described the Middle Ages as being full of beans.) The advantage of legumes is that they fix their own nitrogen from the air, thus being self-fertilizing, and enrich the trace elements in the soil. They also provide a diet rich in protein, or amino acids. With the combination of carbohydrates from cereals, fats from animals, and the 50 per cent increase in output from the three-field system, the population of Europe took off. The way people lived began to change very much for the better.

To begin with, there were more people. This may seem to the modern mind accustomed to the problems of overpopulation to be more of a hindrance than a help. But in medieval Europe, with a population one-tenth what it is today, the scarce commodity was labour, not raw materials. The increase in population meant that more people worked on the land, and that in turn increased output. By the thirteenth century the communities had a surplus to support all kinds of craftsmen who were not producing food: weavers, butchers, smiths, builders, carpenters and so on. They began to produce goods surplus to the needs of their community, and, aided by well-shod horses, trade between village and village grew. The food producers and the craftsmen began to take their profits in money, and to use that money to buy what in earlier times had been regarded as luxuries. All over Europe fairs sprang up to handle this growing exchange of surplus. What one area had too much of, another would lack. Gradually the business of handling the transactions involved was taken over by men who made commerce their craft. As trade increased these merchants would in turn use some of their surplus to embellish the small towns in which fairs were held, spending the money to enhance the status of the town in order to attract more trade to it. Some of the money helped to finance the building of the great medieval cathedrals, and in many of them, notably at Chartres in northern France, the local merchants and tradesmen paid for stained glass windows as memorials of their activities.

By now Sunday was no longer the tired and prayerful respite from backbreaking toil that it had been in the past. With property to manage, business to talk about, taverns to visit, people began to enjoy themselves on their day off. In 1365 Edward III issued a proclamation against 'whoring' in the fields, and against time wasted evilly in pastimes such as dicing, football, dancing and playing games. The real problem for those in power was that these new habits detracted from the one legal activity for a Sunday (besides prayer), which was practice at the archery butts. It took years for a man to be trained to use the six-foot yew bow, and here was the adult male population frittering away valuable training hours enjoying itself. The reason therefore that the longbow enjoyed such a brief spell of success lay in the rising standard of living which the agricultural revolution had brought about. Fortunately for the belligerent royal houses of Europe, by the fourteenth century a new and easier way to kill people had been found: a Chinese way.

Above: Gathering beans in Lombardy during the fourteenth century. Their dietary value was not lost on the people of the time; this illustration comes from a manuscript entitled A Notebook on Health.

Right: Country revels in fifteenth-century France. Peasants enjoy the pleasures of surplus food and spare money, close to the safety of the town where they sell their produce. In the sky an angel celebrates a time of peace and plenty.

66

That the Chinese discovered gunpowder is not in doubt. What *is* in question is whether or not they used it in the same way as the Europeans did. Perhaps the entire question of Chinese invention is worth a brief digression at this point. The major inventions attributed with certainty to the Chinese include paper, silk weaving, clockwork, astronomical instruments, the horizontal loom, the spinning wheel and the waterwheel. These are inventions fundamental in the history of man as a tool-maker. The medieval Chinese were without doubt the most fruitfully inventive people on Earth. However, the fact that the technology of the modern world is Western shows to what extent the two cultures were different at a time vital in the history of the effects of innovation on society. In the stable, civilized East the innovations were not permitted to bring about radical social change as they were in the brawling, dynamic West. The chief reason for this may have been the stultifying effects of Chinese bureaucracy, which owed its origins to the geographical nature of the country. China is a land of wide plains and major rivers. Early in recorded history the Chinese undertook vast irrigation schemes, and the scale on which manpower for these projects was mobilized demanded firm, centralized planning and control. The civil service which evolved to run the irrigation schemes was to remain in power for thousands of years, guarding its position and privilege against change, maintaining a society rigidly stratified into classes between which movement was virtually impossible. There was no drive for the individual to use technology to improve his lot and so rise in the world, because rising in the world was out of the question. Thus it was that invention may have come from the East, but it was only in the West that it brought widespread change. To oversimplify the case: in China gunpowder propelled arrows, and even exploded grenades; in the West it destroyed cities.

The first known recipe for making saltpetre, which is the principal ingredient of gunpowder, was given by the Chinese Wu Ching Tsao Yao in 1044, twenty-two years before the battle of Hastings took place. In Europe saltpetre was found in its natural state in pigsties, cellars or manure trenches, where it was produced by the interaction of bacteria in dung and certain kinds of soil with lime and urine. Dung and earth were often used to make the walls of outbuildings where animals and men would urinate against them, producing the right chemical changes over a period of time. Later on, when nitre pits were developed to produce the saltpetre at will, the urine of a wine-drinking bishop was said to be especially effective. Once the saltpetre was extracted or scraped off the walls, it would be layered together with earth, wood ash and lime. Water was allowed to trickle through the layers, and was then collected at the bottom and boiled with lye and

The first references to gunpowder in Europe were made by the Arab traveller Ibn al Baitar, who said the Egyptians knew of it and called it 'Chinese snow', from the colour of the earth where deposits of saltpetre are found.

The picture shows a saltpetre factory in Germany. The concentrated liquid was first cooled in the tall vat (left) then further concentrated in the boiler on the right. It was left to evaporate in flat pans (top right), from which the crystals were then collected.

alum. These were materials widely in use in the textile industry of the time, as astringents for removing fat and grease from freshly shorn wool and from newly woven cloth. The boiling of the liquid eventually reduced it to a point where crystallization took place. These were crystals of nitre, which when mixed with crushed sulphur and charcoal would explode if ignited. The mixture was gunpowder. By the end of the thirteenth century there were references to gunpowder 'factories' in Germany, and a book was published by a man called Mark the Greek giving the recipe and suggestions for use.

Nobody knows who hit on the idea of using the powder to propel stones through the air from metal tubes, but the first cannons – called bombards – were made paradoxically by craftsmen who had for several centuries been associated with the doctrine of peace: they were bell-founders. The bells were made of bronze, a mixture of copper and tin, easy to smelt and work because of the relatively low temperatures at which the ores melt. (Iron demands higher temperatures if it is to be cast, and cast-iron bombards did not appear until the end of the fourteenth century, when water-powered bellows were used to raise the temperature in blast furnaces to a high enough level.) It may well be that the first bombard was a bell, turned upside down and filled with stones and gunpowder, since the early bombards look very much like upturned bells – although there is one curious manuscript illustration from the end of the thirteenth century that shows an object like a bulbous vase from whose narrow neck an arrow is flying.

The first recorded use of the new bombards to capture a city seems to have been at the small Italian town of Cividale in Friuli, near the modern Jugoslav border. These early bombards were monstrous, crude machines, as dangerous to the bombardier as to his target. More often than not they blew apart. They were also extremely inaccurate, and could only be fired a few times a day, so their main job was to strike fear by show rather than to cause actual death or destruction. Nor could they be moved with ease. In the fourteenth century they were mounted between huge blocks, and dismantled when a siege was over. Their psychological effect, however, was tremendous. Towns would surrender at the mere mention of one of the monsters on its way. But soon enough, as the quality of the gunpowder became more reliable and the stones were replaced with iron balls, castle walls began to fall. By the middle of the fifteenth century no prince could afford to be without his cannon, and demand for the new miracle weapons grew rapidly. Few inventions have ever been seized upon more greedily, and as a result the waging of war became ruinously expensive – not in terms of the assets a prince might need to finance his venture, but in terms of cash with which to pay for it.

Cash was the commodity in shortest supply throughout the Middle Ages, and never more so than in the late fifteenth century, when the Black Death receded. Prices rose in some places as much as 400 per cent, and although many survivors were left with twice the assets they had had before the plague, simply through inheriting what had belonged to those who died, the increase in demand for goods by the newly rich produced a drastic shortage of coin. There was, simply, not enough to go round. Then, in answer to the demand, prospectors began to find silver in the Hartz mountains, and in 1516 one of the greatest silver strikes in history was made. It was in an area of the mountains which now lies within northern Czechoslovakia, near the town of Jachymov, then known as Joachimsthal. The mountain valleys provided two vital aids to mining: water power from the falling streams to run the mining machinery, and wood to build with and to provide the charcoal needed for smelting the ores. A mining boom followed the Joachimsthal discovery, and thousands came to the valleys to seek their fortunes in the mineshafts. At peak output, between 1515 and 1540, Joachimsthal was producing three million ounces of pure silver a year from its 135 veins, and the mint in the town was coining the silver as fast as the stamping machines would go. The coin they minted was called a *joachimsthaler*, shortened to *thaler*, the word from which the modern 'dollar' comes.

A fifteenth-century illumination showing the new bombards. The small culverin is transported on wheels, whereas the larger bombard is crudely mounted on the spot.

This painting from the last decade of the fifteenth century shows life in a silver mining community at Kutná Hora, not far from Joachimsthal. In the foreground the miners work in the early, crudely excavated tunnels. Above them can be seen the activity at the pithead, where the horse capstans haul up the ore which is then washed and taken to the ore market, in the upper part of the picture.

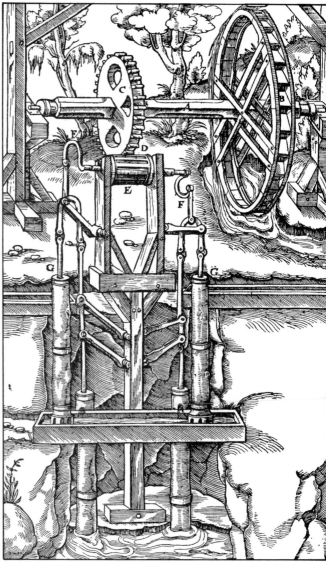

By 1520 the town had a population of over 15,000, and in 1527 a man named Georg Bauer was appointed town physician at the age of thirty-three. Bauer is known to posterity by the pen-name he chose: Agricola. As well as being a qualified doctor, Bauer had a degree in classics and had travelled and studied in Italy, where he had become interested in scientific matters in general. After three years he quit his job in Joachimsthal to devote himself to the problems of mining, and in 1533 he began to work on an analysis of mining in general. It took him seventeen years to finish, and when it was published in 1556 it became the miner's Bible for almost two hundred years. It was called *De Re Metallica* (On the Subject of Metals) and encompassed every facet of mining from geology, to assaying, to smelting, to shaft construction, to drainage. In the section relating to drainage, Bauer put his finger on a problem that was being encountered increasingly all over Europe, in silver mines as well as those sunk to find iron, salt or

Two of the methods used for draining mines, as illustrated in Agricola's De Re Metallica. *Far left, the rag and chain system, successful only against minor flooding. Left, the suction pump, working in 32-foot stages. Note the use of waterwheels to drive the pumps: the rule of the time was 'water must be driven by water'.*

alum. He said: 'One of man's reasons for abandoning mines is the quantity of water which flows in . . . they cannot draw it out with machines because the shafts are too deep.'

Bauer described the three methods, sometimes used separately, sometimes in conjunction, that were failing to drain the deepest mines. All three were powered either by a waterwheel or a horse capstan. One method was to attach porous cloth balls at intervals along a circular chain, reaching down into the flooded area. As the balls passed through the water they would absorb some of it, and as they returned to the top of the flooded shaft they would be squeezed dry, to return to the water later in the cycle. The second method used a giant wooden screw, with one end dipping into the water. As the screw rotated the water ran up the threads to the top of the shaft. The third method attached a crank to the rotating waterwheel shaft. The crank operated a piston inside a cylinder whose open bottom end was in the water. The piston would suck the water up the cylinder, a valve at the bottom would close, one at the top would open, and the descending piston would force the water out of the open valve. This last was by far the most effective drainage method, but it suffered from a mysterious limitation: the piston would not suck water higher than about 32 feet above the surface of the flood level. Although the discovery of the silver mines of South America in the 1540s knocked the bottom out of the European silvermining business, the iron mines were just as vital to the economy in Europe, and the problem of how to drain them continued to be a matter for urgent discussion throughout the remainder of the sixteenth century.

The matter was still unresolved in 1630, when a Genoese called Giovanni Batista Baliani wrote to Galileo asking him to explain the mysterious 32-foot maximum. Galileo did little about the question until a few years later, when, three months before his death in 1641, he was joined in Florence by a young assistant from Faenza, in northern Italy, called Evangelista Torricelli. Torricelli was a mathematician who had studied in Rome, where he had become interested in the subject of the vacuum. Although Galileo had said such a thing could not exist, a few of Torricelli's friends in Rome had been secretly experimenting to see if in fact it could, and when Torricelli heard of Baliani's inquiry he decided to press the matter further. He became convinced that the reason the water would not rise higher than 32 feet had something to do with the weight of the air pressing on the pool of water at the foot of the mineshafts. As mercury was very much denser than water, he realized that its use would save him the bother of having to construct equipment big enough to handle the problem using water, since with the mercury he could scale everything down by fourteen times – the amount by which mercury is denser than water.

Torricelli's experiment which proved the existence both of air pressure and the vacuum. At this early stage laboratory tests involved the use of water, and in every tube of similar diameter the water column stood at the same height.

In June 1644 Torricelli wrote to a colleague and friend in Rome, Michelangelo Ricci, to explain an experiment carried out by his own assistant Vincenzo Viviani, to which he added drawings in the margin. Viviani had filled a 6-foot long tube with mercury and upended it in a dish full of the same metal, with the open end of the tube beneath the surface of the mercury in the dish. When he took his finger away from the open end of the tube, the mercury in it ran out into the dish, but stopped when the mercury column left in the tube was still about 30 inches above the dish. Torricelli reasoned that the weight of the air pressing on the mercury in the dish had to be exactly equal to the weight of the mercury left in the tube. If there were no weight of air, all the mercury in the tube would have run out into the dish. If this were so, he wrote, 'we live submerged at the bottom of an ocean of air'. What was more, in the space at the top of the tube left by the mercury was the thing thought to be impossible: a vacuum. Torricelli wrote that if he was right, the pressure of the air in our atmospheric ocean must vary according to how far up or down in the ocean we were.

Ricci, realizing that current Church opinion in Rome would not take kindly to these arguments (since if they were true several things followed, such as the existence of interplanetary vacuum, with sun-orbiting planets), made a copy of Torricelli's letter and sent it to a priest in Paris, Father Marin Mersenne. This man was an extraordinary Minorite friar who ran a kind of scientific salon, to which came many of the more radical thinkers of the day. Following his habit of copying letters he received and circulating them among his many scientific contacts throughout Europe, Mersenne became known as the postbox of Europe. It was precisely for this reason that the copy of Torricelli's letter ended up in Mersenne's hands, and sure enough the first thing he did was send another copy of it to a friend who was interested in the same problem, the son of a Paris tax inspector, Blaise Pascal. Two years after receiving the letter (he had been busy meanwhile in the Paris gambling halls working on laws-of-chance mathematics) Pascal found himself in Rouen. It was here that he repeated Torricelli's experiment, to check it; only he did it full-scale, with water. Unfortunately he was in no position to check the second part of the argument – there were no mountains around Rouen. However, Pascal had a brother-in-law called François Périer who lived in central France, in Clermont Ferrand, which is surrounded by mountains. So Pascal wrote to Périer, asking him to take things to the next stage.

On 19 September 1648, one year after Torricelli's death, Périer and a few trustworthy friends (clerics and town councillors) left one tube of mercury upended in a dish at the bottom of the Puy de Dome, and slowly climbed the mountain with a similar set of equipment. At the summit, 4000 feet up, the other tube was upended and examined

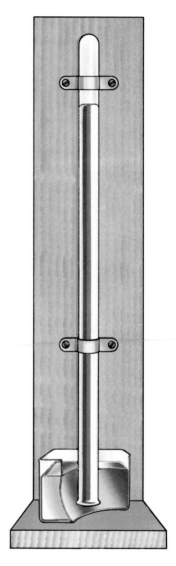

The basis of the modern barometer. As the atmospheric pressure changes with varying meteorological conditions, the level of mercury rises and falls. Good weather means high pressure, bad weather, low.

in various places and in various conditions: inside the ruins of the Roman Temple of Mercury, outside it, in wind, sheltered from wind, in fog, in sunlight, in rain. Nothing altered the fact that, as Torricelli's letter had predicted, the level of the column of mercury in the tube was lower than that of the one they had left at the foot of the mountain. They concluded that the air pressure, being less at the top of the mountain, would support less mercury in the tube. Everyone was elated. The barometer had been invented.

The experiment on top of the mountain represents another of those moments that occur in the process of change, in which things come to a nexus. The invention of the barometer and the discovery of air pressure suddenly multiplied the number of possible routes that the path of innovation could take. The event led to investigation of the behaviour of gases, the discovery of oxygen and the development of respiratory medicine. It also led to hot-air balloons and the jet engine. It led, through the interest in how light rays–and later other types of ray–behaved in gases, to the cathode ray tube and to radar. One of the oddest routes was taken because of a French astronomer coming home late one night in 1675, happily swinging his barometer. M. Jean Picard, returning from the Paris Observatory, suddenly noticed that his barometer had started to glow. The more he shook it, the more light it produced. Picard's experience caused great excitement, and interest grew in this 'glow of life', as it was soon called. Of the many experimenters trying to explain what was happening, Francis Hauksbee, an Englishman and pupil of the great Newton, was the first to reproduce the glow at will. He made tubes with valves in them so as to measure the amount of air let into the tube, and showed that the glow was greatest when the barometer was half-filled with air. In 1705 he told the Royal Society that the glow was caused by friction between the glass and the mercury, and that other materials, such as amber and woollen stuffs, would glow for the same reason.

Hauksbee's Influence Machine. Note the valve on the left of the globe which was opened to permit calculated amounts of air to enter the evacuated glass.

A year later, in 1706, Hauksbee produced his Influence Machine. By cranking round a large drive wheel he was able to spin a glass globe inside which a vacuum had been produced. A valve in the globe permitted the gradual reintroduction of air, and he demonstrated that when the globe was half-filled with air, the mysterious 'luminosity' could be produced at its brightest by pressing a hand lightly against the spinning globe. The crackling sound this produced reminded him, he said, of lightning. The spinning globe also attracted various metal flakes, threads, and so on. It was time for further investigation, Hauksbee suggested. In 1729 Stephen Gray discovered that if he rubbed hard at a glass tube with a cork in one end of it, a thread attached to the cord would carry the attractive force down its length and cause a feather to stick to the other end, at a distance of nearly 800 feet.

In the years that followed, the investigation of electricity took on the air of a fairground; itinerant lecturers tramped around Europe with cartloads of equipment, astonishing the crowds with glow and attraction alike. One of the most bizarre results of this craze for the new force was the establishment in London, by James Graham, of the Temple of Health. The principal attraction was a great 'celestial or magnetico-electrico' bed, in which childless couples might indulge in reproductive activity under the influence of the therapeutic electric field created by a Hauksbee machine. One of the ladies who acted as 'goddess of Health and Hymen' on these occasions was no less a personage than Emma Lyon, later to become famous as Lady Hamilton, Nelson's mistress. An experimenter in Germany, called Christian Hausen, caused great excitement by suspending small boys in a wooden frame, using them as conductors to see whether they would attract materials when connected to an Electrical Machine.

In 1745 matters were advanced by a Pomeranian clergyman called Ewald Jürgens von Kleist and a Dutch Professor of Physics called Petrus van Mussehenbroeck, who were simultaneously attempting to store electricity in a glass jar full of water by 'filling' it with electricity from a Hauksbee-type machine. Sure enough, after the water was 'charged', they touched it and received a considerable shock. The jar was developed at the Dutch University of Leyden; it became known as the Leyden jar, and spurred further work. In 1746 the French abbot

A lecture on electricity at the Amsterdam society for adult education. While Professor van Swinden points out the electric flash between glass tubes, his assistant cranks an electrostatic generator, and a spectator studies the new Leyden jars to the left, behind the blackboard.

Jean-Antoine Nollet attached a circle of monks, all holding hands, to the jar and watched them jump from the shock as electricity ran from one body to the next. Nollet enjoyed the privileges of the experimenter in a way that was particularly French. As one of his aristocratic lady admirers wrote, 'The Abbot writes to me that only the carriages of duchesses, peers and beautiful women are lined up outside his door. Here, evidently, is sound philosophy which is going to make its name in Paris. God grant that it may last!'

By 1749 the new force was being used to fire mines and explode gunpowder. Then, in 1786, came the extraordinary news that electricity flowed in animals. Luigi Galvani, Professor of Anatomy at the Italian University of Bologna, had discovered that touching a frog's leg with metal would cause the muscles in the leg to twitch. Ten years later another Italian, Alessandro Volta, showed that the current causing the twitch was due to the dissimilar metals Galvani had used. Volta piled alternate slices of copper and zinc on top of each other in water to which a small amount of acid had been added, and found that the pile of metal sandwiches produced a continuous flow of electricity. In 1800 he published his results, and the Voltaic Pile went on to the market – the world's first battery. As interest in the force continued to mount, people began to re-examine the connection between electricity and the attractive force it seemed to have. Early in the nineteenth century two Frenchmen floated in water a Voltaic Pile made up of 1480 plates, and gloomily noted that the pile did not align itself north–south, as it should have done if there were a connection between electricity and magnetism. The fact that a steel bar close by *did* had no significance for them! Then, in 1820, a Dane called H. C. Oersted was ending a lecture in Copenhagen with an experiment to show that there was no connection between electricity and magnetism.

The craze for therapeutic electricity brought patients to the quacks in hundreds. There was even a suggestion in eighteenth-century England that electrifying the poor would alleviate their misery.

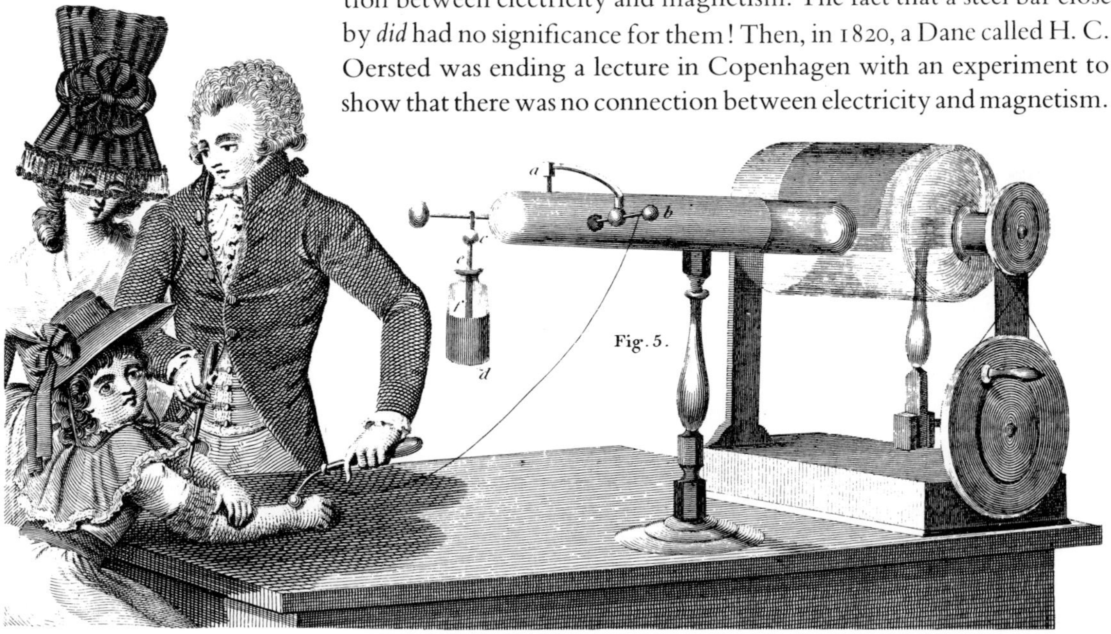

Fig. 5.

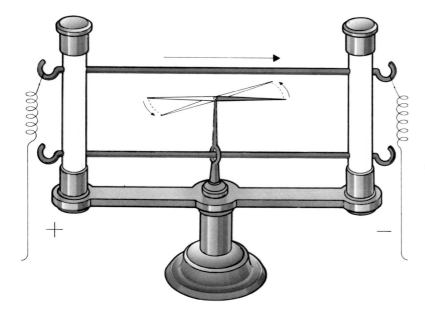

The beginnings of modern electric power. Oersted proved that an electric current flowing in the direction of the upper arrow would set up a magnetic field which would cause a compass needle to point to the poles of that field. This was firm evidence – after two hundred years of investigation – of the link between electricity and magnetism.

He was to show that an electric current passing down a wire would not affect a nearby compass needle – when, to his chagrin, it did. The needle swung as soon as the current was switched on. Oersted then rightly concluded that if the current affected the needle, the current had in some way to have a magnetic field round it.

It was here, in a Copenhagen lecture room, that one of the modern world's most important and influential inventions was born, though Oersted did not know it. In the same year the Frenchman André Ampère, talking of Oersted's discovery, said that the current must be producing the magnetic field. Sure enough in 1825 an Englishman called William Sturgeon wound a live wire round a bar of soft iron, and in doing so made an electromagnet. Then Michael Faraday did it the other way round, putting a spinning magnet inside the arms of a horseshoe-shaped coil of wire, causing it to become electrified. Then, in 1857, a German scientist called Hermann von Helmholtz, a pianist who was interested in how the ear worked, discovered that he could make his piano strings vibrate by singing into his piano; he used an electromagnet, switching on and off, to attract the arms of a tuning fork, causing it to vibrate and produce sound.

The work of all these men was brought together, in Boston, by a Scotsman who had been trying to teach deaf-mutes to talk. In 1874 he saw a phonautograph at work for the first time in the Massachusetts Institute of Technology. The device had been developed by a Frenchman called Leon Scott, who attached a thin stick to a membrane. On the other end of the stick he placed a bristle, set to touch a piece of smoked glass. When he spoke into the membrane, itself placed on the narrow end of a cone to concentrate the sound, the membrane

The Leyden jar – the first electric battery. The jar, filled with water, could be charged with static electricity through contact with the vertical rod, and discharged when contact was made between the rod and the metallic coating on the outside of the jar.

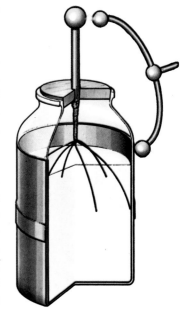

vibrated, and the bristle produced a wavy line on the smoked glass. What was more, the pattern drawn was related to the sound. Each vowel produced a distinctive pattern. The Scotsman was delighted. 'If we can find the definite shape due to each sound,' he said, 'what an assistance in teaching the deaf and dumb.' He had previously been using a tuning fork to teach his pupils to feel the vibration created by sound, and his contacts at M.I.T. had taught him the theory behind the work of Oersted, Sturgeon, Faraday and Helmholtz. In 1875 he put their ideas together in one single system. He wound wire round an iron bar, connecting one end of the wire to a battery, and the other to a second bar at some distance away. Close to each bar he placed a metal foil membrane fitted on the narrow, open end of a cone. As noise caused one of the membranes to vibrate, the metal moved back and forward in the magnetic field round the wire, causing the field to fluctuate. This in turn caused current to fluctuate along the wire to the second electromagnet, setting up in it another fluctuating magnetic field, which attracted the second membrane, causing it to vibrate in exactly the same pattern as the first. As it vibrated it produced sound. In this way, the sound which caused the first membrane to vibrate was reproduced by the second membrane.

The telephone, like so many inventions, was the synthesis of the work of many other people.

The Scotsman who put everybody else's work together in a brilliant synthesis was called Alexander Graham Bell, and his new sound-reproducing machine was the telephone.

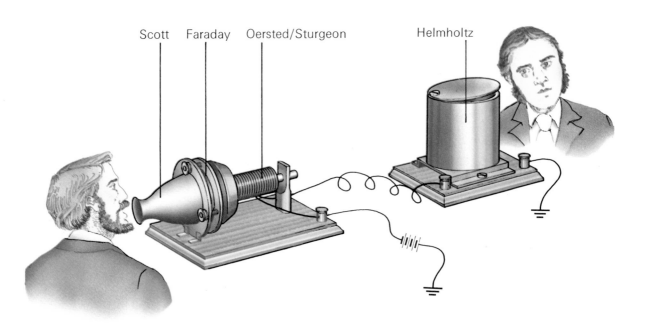

4
Faith in Numbers

Telecommunication, born of Oersted's discovery of the relationship between electric current and magnetic field, has had the most profound effect on the development of the modern world. Apart from the effect of the telephone, television and the computer database on the individual's personal life, telecommunication has enabled the modern nation to organize itself in a way that would have been impossible before. The economic, political and cultural life of a country is shaped by the way it uses its communications systems to encourage or deny access to the information which such systems transmit. At an international level the ability to exchange vast amounts of information almost instantaneously has made possible the operation of giant, multinational corporations whose existence blurs the distinction between national boundaries, draws the countries in which they function closer together, and calls into question the old concept of national sovereignty. It has also made possible the existence of international political bodies which can sit in permanent session only because they are able to maintain continuous contact with their individual governments. Argument has become a substitute for conflict.

   In a world whose interdependence has been fostered by telecommunication, the ability to make instant contact with almost any place on Earth has enabled us to tune our economic and political activities to a fine pitch. And yet we are now on the threshold of a revolution in communications technology of immense significance.

A 1470 miniature illustrating activity in the Cloth Merchants' Street, Bologna. Cloth making wa. one of the major elements contributing to the economic growth of Europe in the late Middle Ages. Note the tailor (centre) measuring a prospective client. By this time the choice of cloth and style was so varied that laws had been passed to determine how each social class was supposed to dress.

Technology has placed personal communications devices within the reach of every individual. Thanks to direct broadcast satellites, the impact of Western (and in particular, American) culture is now felt directly by countries whose religious or ideological beliefs may be threatened by the intrusion. We live in a post–cold war, post-imperial world, made up of the heterogeneous fragments of the nineteenth-century empires of France, Britain, Spain, Portugal, and Holland. These fragments at present are held together by the telecommunication network and share the knowledge each member chooses to transmit on that network. As more nations obtain the technology necessary to build and maintain their own networks, are we facing a period of informational conflict? And how will such a development affect the loose confederation of our post-imperial world?

The last time a world empire fell apart it was Roman, and there, too, the fragments of the Roman province of Gaul, which had broken up into several small kingdoms, were held together by a communications network which preserved some of the imperial administration techniques. The operators of the network also guarded the most advanced technology of Rome, and ultimately it was this technology which was to help relocate the centre of power in Europe northwards away from the Mediterranean in the tenth century. One of the greatest examples of that technology in the Roman world was the huge imperial grain mill at Barbegal, near Arles, in southern France. It was erected in the third century A.D., and may have been the largest industrial complex in the Roman Empire. The mill consisted of eight pairs of waterwheels, set at intervals down the side of a 65-foot slope. The wheels were powered by water falling from a reservoir at the top, which in turn was fed by an aqueduct built for the purpose. The sixteen wheels each powered two grindstones, through a gearing system that turned the horizontally revolving shaft of the waterwheel into a vertical one, on which the grindstones were set. The mill ground flour from as far away as Egypt, and though the local provincial capital of Arles had a population of 10,000 the Barbegal mill produced enough flour to feed eight times that number, leaving a surplus for export, or to provision the legions garrisoned in Gaul.

At almost the same time as the Barbegal mill was being built, the Roman Empire was splitting into two halves, one administered in Rome, the other in Constantinople, the new city built by Constantine on the Bosporus. From this point on, Rome had to support herself without the wealth of the eastern part of the Empire on which she had previously been able to draw. Yet the vast bureaucracy which that wealth had spawned and the 300,000-strong army it had funded were both still there, and the only way to support them was by raising taxes.

Telecommunications, using satellite stations like this one in Kuwait, make it possible to dial a telephone number anywhere on Earth by bouncing radio waves off satellites. These are set in orbit 23,000 miles above Earth, where they circle the planet at the same speed as the planet rotates, thus remaining over the same spot. Because of this, once transmitting and receiving antennae have been aimed at the satellite they are unlikely to need repositioning.

So began a chain of events which was to lead to the fall of the Western Roman Empire within less than two hundred years, destroyed by its own taxation system. Higher taxes devolved on to tenants of either land or buildings as higher rents. After a time the tenants had less surplus with which to support their families, so the birth-rate fell. At the same time, administration and collection of the new taxes demanded more bureaucrats, and in order to support them taxes had to rise again, so the population declined further. It was this descending spiral that ruined the West, as the economy faltered and began to grind to a halt.

In the fifth century, as the legions began to withdraw to protect Rome, the Germanic tribes which had been in contact with the Empire for over two hundred years gradually consolidated their position. In the province of Gaul they had held high administrative positions since the fourth century. When the so-called barbarians invaded in the fifth and sixth centuries they were fighting Romanized Franks or Burgundians, not Romans. And with the armies gone and the local populations so long forbidden to carry arms, resistance was apathetic. Small city-states sprang up. The great estates, established as part of the imperial economic structure, had no further *raison d'être*, and they gradually ran down. The imperial roads were too expensive to keep in repair when there were no legions to use them. They served no local purpose, and life had become local, so they too fell into disrepair. Economic activity dropped sharply as the province split into tiny self-sufficient units under local kings, and especially during the plagues of the sixth and seventh centuries when the population of Europe was halved.

The network of communications that maintained contact between one part of this patchwork quilt of territories and another was that of the Church. By the fifth century the diocesan organization of the ecclesiastical hierarchy corresponded to the concentrations of civil population. When the legions withdrew, the administration of the area fell into the hands of the bishops and their clergy: they could read and write and the new rulers could not. For this reason the Church was granted many privileges, in particular exemption from taxes, that helped it to survive, while the Church itself exacted a tax of one-tenth (a tithe) from its own tenants. By the eighth century Europe was scattered with churches and monasteries, many of which had to provide a service that no one civil community could have done, in the absence of a centralized power: they ran the mails. A new church or monastery was called upon to provide pack horses or messengers, and in some cases a freight service of wagons, within a radius of up to 150 miles from the church. It would seem that the Church had a bishop-to-bishop communications network that continued to operate right

This carved ivory book-cover from the tenth century shows St Gregory and his scribes at work, making copies of manuscripts. Such copies were needed to replace worn-out originals and to stock the libraries of the monasteries which were spreading throughout Europe at the time.

through the Dark Ages, connecting one kingdom with another, carrying news and information as well as ecclesiastical business, and transmitting knowledge in the form of copies of manuscripts.

It was in the writing departments of their monasteries that the churches preserved the few elements of technology that would otherwise have been lost to barbarians who had no use for mills on the scale of Barbegal, for instance. The earliest description of the Hellenistic gearing system employed by the mills is in the writings of a Roman engineer, Vitruvius, who referred to them in 14 B.C. as 'machines that are rarely used'. The ancient world made little use of the waterwheel – the power source for the mill – for industrial purposes. There are various reasons put forward, the one most generally offered being that with a high slave population the Greeks and Romans had little use for labour-saving devices. It may also have been that in a centralized imperial administration, large areas of the country could be supplied from few central sources such as the mills at Barbegal, and it was only when communities became separated and needed to survive that each one built what machinery it needed to live on. Certainly all through the period from the fifth to the tenth century there are constant references to mills. In the main these references are found in law, such as that passed by Theodoric the Goth, in sixth-century northern Italy, to prevent the diverting of water supplies 'for private milling'.

There were two basic types of drive for the mills: the overshot wheel, powered by water falling on the wheel paddles from above, and the undershot wheel, driven by a current moving against the paddles at the bottom of the wheel. Which one was used obviously depended on whether the water supply was running flat, like a river or mill pond, or falling, as it would on the side of a steep slope. The gearing system referred to by Vitruvius gave the water mill its immense flexibility. With gears the mill could operate on a vertical wheel, which could be placed in or under a water source little wider than the wheel itself. The narrowest streams would provide enough water. Gearing also allowed manipulation of the ratio between the speed of the wheel and the rate at which the grindstones turned, so that the rate of water-flow was not a limiting factor in the siting of a mill. A fast-flowing river could be slowed by means of the gears; a slow stream could be accelerated.

The extent to which the mills were used from the fifth to the tenth century was also related to the profit that could be made by a mill owner. In most cases such owners were churchmen, since it was they who had the knowledge to construct the mills, and the literacy to work out accounting systems to run them as businesses. They would lease the mills for given periods to farmers, taking the profit in flour rather than coin, since coin was then a commodity few people saw in a lifetime.

The great waterwheel. Vitruvian gearing enabled the power of the turning wheel to be moved through the vertical plane and back to horizontal, allowing the millstones above to be offset as the location of the building demanded.

The Saxon word for aristocrat is 'lord', meaning 'loaf giver'. The grain mill had already spread to Gaul (modern France, Belgium and the Low Countries) by Roman times. By the end of the tenth century it was widely used all over Europe. In 1089 the Domesday Book – that nation-wide inventory taken by the Normans after they had conquered England in 1066 – showed that among the possessions they had in their newly acquired territory were nearly six thousand grain mills, in over three thousand locations.

In general, the mills at that time differed little from those described by Vitruvius. But in the ninth century a breakthrough occurred. It may have come about because there was a superabundance of powerful mountain streams in the area, or it may have been born of having to live in mountainous, marginal land with a low labour reserve. It may even have been that a monk came across the idea in a manuscript he was translating from Greek. The device first appears in Switzerland, and probably originated somewhere in the Alps. The first reference to it appears in A.D. 890 in the monastery of St Gall, where it was being used to make beer for the monks. The device in question was the cam, a Hellenistic invention dating from the second or third century before Christ. In its simplest form it is a piece of wood set on the side of a shaft, so that as the shaft turns, the protruding piece of wood strikes against anything placed in its way. The coupling of the cam with the

The overshot wheel, powered by falling streams, permitted industrial operations in the mountain valleys from which came the metal ores to be worked. Here the cams on the driveshaft operate a trip hammer for crushing the ore, and a bellows for use in the blast furnace where the metal would be smelted.

Ioan. Stradanus inuent.   Ioan. Galle excud.

B s f.

The medieval equivalent of the modern power-station, the waterwheel was to be seen wherever there was water. The three major uses to which water power was put are seen here.
Above: A number of grain mills situated in a broad river. The ones on the right of the picture are built out from the bank to benefit from the faster current. In the foreground mill the gearing mechanism is set directly against the wheel.
Bottom left: Water power drives cranks to push levers operating blast furnace bellows and a trip hammer.
Bottom right: The trip hammer pounds newly woven cloth to matt the fibres together, giving the cloth a 'full' appearance.

waterwheel gave Europe the power source it needed almost exactly at the right time, as the Viking invasions in the north and the Arab incursions in the south began to tail off. As peace extended, the population began to rise, the forests were cleared, food production increased, trade began to spread outside the villages to other local communities, and the new water-power technology began to find opportunities for use.

What followed between the tenth and the fourteenth centuries has come to be known as the Medieval Industrial Revolution. The cam was used in various ways: to trip a hammer every time it rotated to its original position, to push down on one end of a lever and activate a suction pump at the other end for raising water from wells, or to act on a crank to turn the rotary motion into horizontal back-and-forward motion for operating knives or saws. The beer-making mill at St Gall used the hammer, to crush malt for beer. A hundred years later, in A.D. 990, there were hemp mills in southern France, hammering rotted flax stalks or hemp to knock out the fibres in preparation for the making of cord or linen. By the eleventh century there were forge hammers in Bavaria, oil and silk mills in Italy; by the twelfth century there were sugar-cane crushers in Sicily, tanning mills pounding leather in France, water-powered grinding stones for sharpening and polishing arms in Normandy, ore-crushing mills in Austria. From then on the use of water power spread to almost every conceivable craft: lathes, wire-making, coin-producing, metal-slitting, sawmills and – perhaps most important of all – in Liège, northern France, in 1348 the first water-powered bellows providing the draught for a blast furnace.

But if the recovery of Europe owed its success to any one particular variant of the waterwheel and cam, it was to its use in the textile industry, in the form of fulling mills. These used the trip hammer to pound alum and other astringent materials needed to clean and take the grease out of newly woven cloth, and then to pound the cloth itself so as to soften and intermix the fibres, giving it a 'fuller' appearance. By the thirteenth century, fulling mills were turning out ever-increasing amounts of cloth in northern Italy, Flanders, along the banks of the Rhine in Germany, and in England. And the source of their raw material – wool – was the same as that of the technology itself had been: the monks, and in particular, an order known as the Cistercians.

To understand why the Cistercians exerted so profound an influence on medieval Europe it is necessary to go back briefly to central Italy in the late fifth century, and to a priest who saw the need to separate and preserve the life of the religious from the degeneracy and confusion he saw around him as the Western Roman Empire went down.

This priest, who lived from A.D. 480 to 543 and who later became known as St Benedict, formulated a Rule by which the monks of his monastery would live. It laid stress on the equal value of prayer, study and work, and in this way Benedict laid the foundations for self-sufficiency in a period when a community would either survive on its own, or not survive at all. At the core of the Rule was the edict, *laborare est orare* (to work is to pray). Benedict's monks were to be no mere ritualistic bookworms – he wanted them to get dirt under their nails. As the movement spread across Europe the Benedictines set up abbeys that prospered, safe behind their massive walls, even when the depredations of the barbarian invaders were taking their greatest toll of the country around. It may even be that the first of the new towns that began to rise in the ninth and tenth centuries started life at the walls of the abbeys, with the regular markets that the monks held on feast days. But the very success of the Benedictine rule was, for a time, to prove its undoing. By the eleventh century the monks had grown fat and rich from their estates. The simplicity of their daily ritual had given way to elaborate services and ornate churches and vestments.

This eighth-century illumination shows St Matthew copying and decorating manuscripts. The contemplative life of the monasteries gradually gave way to political and financial involvement with the outside world, and the Cistercians, founded in 1098, attempted to return to the style of life depicted here.

IN MILI BEAT

Many of the pages of the early Cistercian manuscripts show the Order hard at work in the fields and in their foundries and mills.

VI CON TRA

VERITATIS VERBA

It was a reaction to this decadence that led a monk called Robert of Molesme, in France, to leave the rich Burgundian house with twenty of his fellow revolutionaries in 1098 and set up a rival concern in a poor, marshy part of the Burgundian forests known as Cîteaux – after which they called themselves Cistercians. Thanks to their observation and extension of the original Rule of St Benedict, they were to have a profound influence on the world around them less than a hundred years after the Cîteaux foundation. Several vital additions were made to the Rule by the third Cistercian abbot, an Englishman called Stephen Harding, in a constitution that gave the Order a dynamism that was to make it the most advanced technological community in Europe. The most influential decree was that the foundations of new abbeys were to be 'far from the haunts of men'. This was intended to remove the brothers from the taint of urban life, but it also had the effect of making survival more difficult, on marginal land for the most part at the higher end of uninhabited valleys. A typical site was described in 1203 by a monk of Kirkstall, in Yorkshire: 'A place uninhabited for all the centuries back, thick set with thorns, lying between the slopes of mountains and among rocks jutting out on both sides, fit rather to be the lair of wild beasts than the home of human beings.' If under these conditions a house was also obliged to be self-sufficient, the small number of monks present could not produce the necessary food, so the constitution permitted the introduction of 'lay' brothers, who were in effect little more than farm labourers. Each house managed its own affairs, though at an annual meeting ideas were exchanged on improved methods of operation and management (particularly valuable as the Order's land-holding grew). Above all, a house might sell any excess produce to the communities around it.

With a regular day of four hours of prayer, four of reading and meditation, and six of manual work, plus the ready water-supply that was to be found in most of the abbey sites, placed as they were in foothills, the Cistercians rapidly became masters at making marginal land productive. The Order spread with equal rapidity. By the end of the twelfth century, 102 years after they had begun, there were 530 houses all over Europe, each one of them a medieval factory. Whenever possible houses were built with the local water-supply running through the centre of the site – to provide water for hygiene, and to power the machines the Cistercians became so adept at using. By the 1300s they had foundries with associated mills for treating ore, fulling mills, corn mills, water-powered workshops where tools were made and wool treated, with forges, oil mills, wine presses (the Cistercians set up the great vineyards of Clos de Vougeot) and all the equipment and administrative organization to run the vast business concerns which many of the abbeys had become.

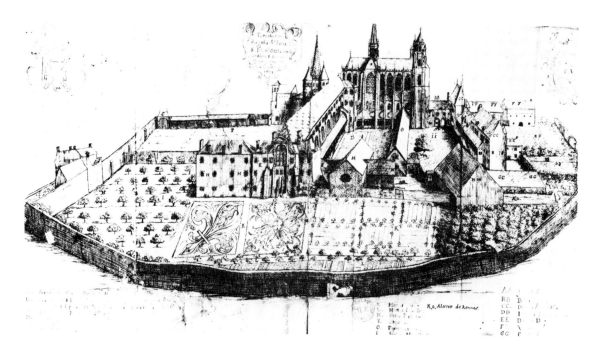

As the Order grew, large grants of land were often made by local aristocrats. The land was usually of the same poor quality as that on which the abbeys themselves stood, so the noble donor could save his soul at little cost. Often these lands were also a good distance from the abbeys, so in the twelfth century 'granges' were set up; these were semi-independent farms which in time became too big or too prosperous to be run from the mother house, and were leased out to tenant farmers. As might be expected, the leases on these properties read like treatises on farming and animal husbandry. The Cistercians had become Europe's best land managers, and they insisted that their knowledge be put to use. The leases contain instructions on irrigation, mill use and upkeep, husbandry, land clearance, crop rotation, management of finance. In this way the experience which the Order had built up over the decades went, so to speak, on the open market, and the laity were quick to copy. The Cistercians acted as any major corporation might be expected to act. If villages were included in a land grant, they were destroyed and the people resettled elsewhere. The Order opened warehouses and finance offices at the major seaports, to facilitate the export of the commodity for which they had become famous by the thirteenth century: wool. As has been said, much of the land they owned was marginal, and what they could not till they used for grazing sheep. Thanks to careful management and expertise Cistercian wool was the best available, and the textile centres—in particular those of Flanders—would buy every pound the Cistercians could produce.

The abbey of St Bénigne, at Dijon – the result of a land grant in A.D. 587. Successful management turned this and other houses like it into rich, self-contained communities, cut off from the encroaching civil population by the mandatory wall.

Flanders had been a centre of cloth production from the tenth century onwards. Until early in the next century the weaving of wool was a laborious process. The loom used was a vertical one, with the warp threads hanging, weighted, and the weaver standing to pass the weft threads back and forth between the warp threads by hand, using a stick called a 'swordbeater' to beat the weft down firmly into place. It was slow, tiring work, and production was consequently limited. Then, at some time in the eleventh century came one of those accidents that are the engines of change. Somebody arrived in Europe, probably from Muslim Spain, with a new kind of loom. It may have come originally from China, brought across the Indian Ocean by Arab traders. It was a radically new design, and it boosted production enormously. On this new loom the threads were stretched horizontally on a frame. Two horizontal boards above the frame each supported two more horizontal boards, with holes in them. Through the holes in one board passed the even-numbered threads of the warp, while the odd-numbered threads passed through holes in the other. These boards were lifted alternately by the use of foot pedals attached to the overhead supporting boards. All the weaver had to do in order to pass the weft thread through alternate warp threads was to press down on the pedals, and the threads would lift, leaving a gap through which the weft thread could be thrown, attached to a shuttle. On the return journey the alternate threads would be lifted, and so on. The swordbeater for packing the warp now came in the form of a permanently attached comb-like device, pulled back towards the weaver whenever necessary. The speed of weaving on this loom was much greater than on the old vertical one, and production rose fast.

The machines that revolutionized textile production in the fourteenth century: the horizontal loom and the spinning wheel. Note the design of the shuttle which carried the weft thread, virtually unchanged today, and the balls of wool ready under the loom.

This was the device that made Flanders rich, because it made a fine woollen cloth which when finished had an almost silky feel to it. However, the new loom created a problem because it considerably increased the demand for thread, and at the time this was still being spun by hand. The wool was teased out of a fluffy mass, and twisted as it emerged, tied to a bobbin that both acted as a weight to keep the fibres taut, and as a core on which the thread would be rolled. It was slow work, too slow for the new loom. The textile industry was stuck with this bottleneck in production, so when the first overland expeditions left Europe for the East, people went looking for new materials and new technology. It may have been one of these who returned with the answer to the problem: the spinning wheel. The first European reference to it is in Speyer on the Rhine, in 1280, where it is described as the 'great wheel'. It spun thread by using a large wheel, either turned by hand or through the use of a foot pedal and crank; with a system of pulleys this drive wheel caused a small hollow spindle to turn. The mass of wool was stuck on a rod attached to the frame, and the teased wool went through the spindle, being twisted and pulled taut as it did so, and was then wound on a bobbin. The spinning wheel and the horizontal loom fitted together like pieces of a jigsaw. The result was a tenfold increase in cloth production.

The money that the new technology made for the burghers of Flanders was a legend in medieval Europe. John Gower, a fourteenth-century contemporary of Chaucer, writing in his *Speculum Meditantis*, says: 'O wool, noble dame, you are the goddess of merchants. To serve you they are all ready. You make some mount to the heights of riches and fortune, and you cause others to fall to ruin . . . you are cherished throughout the world . . . all over the world you are taken, by land and sea, but you are directed to the richest people.' The towns of Flanders used the profits brought by wool to support musicians,

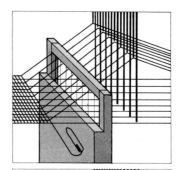

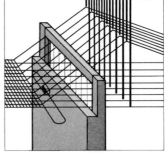

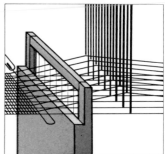

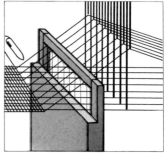

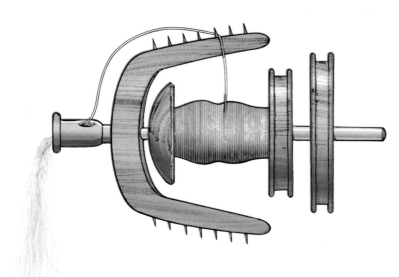

Left: The thread maker, driven by the spinning wheel. The fibres were twisted by the rotation of the toothed flyer, turned by the larger pulley wheel (right). The bobbin, turned by the smaller wheel, spun slightly faster, putting tension on the thread.

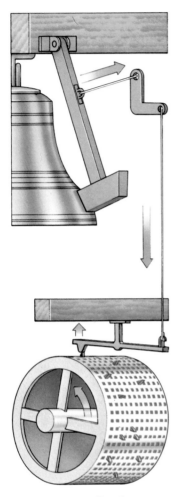

painters and writers; the display of wealth extended too into people's homes and into their dress. Once a year at the wool festival in Bruges the merchants and workers paraded, their banners proclaiming the power of the loom and the spinning wheel. The same technology that powered their factories and paid for their luxuries went into their churches to save their souls. The famous carillon of Mechelen – a more complex peal than that heard anywhere else in Europe – rang the bells on feast days, operated by the same cam that worked the trip hammers of the fulling mill.

As peace descended on the north with the failure of the Scandinavian and Arab invasions, tenth-century Europe began to stir. With travel becoming safer there was a natural urge for a community with surplus to move it to areas where there was deficit, in exchange for profit. In the beginning the commodities on the move were grain from Picardy and Normandy, wine from central and eastern France, wool from Flanders, the Auvergne and Cévennes, and flax from the north-west of France. These moved along the ancient routes established by the Romans, east–west and north–south, and in the first area in Europe to settle after the end of the tenth-century invasions – the region of Champagne – the crossing-points of those trade routes became markets. The first we hear of them is in 1114, when already there were regular fairs held in succession through the years in four towns of Champagne: Troyes, Provins, Lagny and Bar-sur-Aube. The main fairs seem to have been centred on Troyes (summer and winter) and Provins (May and autumn), and at their height these very first European-wide markets, known as the Champagne Fairs, attracted merchants from extraordinary distances, when one considers that there were few if any roads, no protected routes along which trade moved on wagons or pack-horses, when a man took his life in his hands to move far from his own town at night, when the forests were still so thick as to be impenetrable. Yet come they did, from Flanders with cloth, from Lucca in Italy with silk, from Spain, North Africa and Provence with leather, from Germany with furs and leather. Luxuries from the East were brought through Venice: spices, wax, sugar, alum, lacquer, and from all over Europe came horses, livestock, jewellery, grain, wine, dye-woods, cotton.

The Milanese were the first to arrive, in 1172, looking for wool. (It must be remembered that wool was an expensive luxury at the time – unlike linen, which in its raw form of flax was cheaper to produce than sheep.) After the Milanese came the Genoese, the Pisans, the Venetians and those from other Italian cities; while the Fairs lasted, the Italians were always the most numerous foreign group. It was the Italians who invented the *commenda*, a piece of paper that first made international trade possible – granted safety on the way. It detailed an

Above: The carillon shown operating a bell. This cylinder could ring twelve bells, the operating lever for each bell being placed above one of the twelve lines of holes around the cylinder. Only one peg was needed in each line of holes to make all twelve bells strike once per revolution.

Above left: How the process of weaving is carried out.

arrangement between investors, who would stay at home and take three-quarters of the profit, and merchants, who would put up the other quarter and do the travelling. The *commenda* gave the merchant *carte blanche* on what to do with the goods and where to go to sell them. The only proviso was that on his return he was obliged to make a detailed and fair statement of profit and loss to his investors. The incentive to honesty was evident enough. No merchant who was suspected of trickery would travel again, since nobody would back him. The *commenda* sent the merchants to Champagne in their hundreds. The Fairs had their own administration to handle the crowds. There were wardens of the peace, sergeants of justice, clerks to collect taxes on residence and stall-holding, on entry and exit tolls and on sales taxes, and to organize safe-conducts for Jews and Italians. Each country or city-state represented there had its own consulate to handle disputes between fellow countrymen. The Champagne Fairs were special events, unparalleled in European history. Of course, as in later times, there would have been the usual riches, the crowds from strange places, the brawling and foreign tongues and stories of distant adventures. But at Champagne there must have been a sense of *déjà vu*, of the dimly remembered past when Europe had been one, a feeling that the Fairs were a coming-together again, as memories of the anarchy and terror of the decades of invasion receded.

Left: A fourteenth-century medieval fair held under the patronage of the local bishop. Towards the back of the picture are temporary stalls set up by the visiting merchants.

The great thirteenth-century financial revolution. The Fairs gave rise to money-changing as merchants traded in different currencies. The banks, or counters, where this took place gave their name to the financial institutions that followed. Note the book containing rates of exchange.

With so much money travelling such long distances it is inevitable that the Fairs should have become the first international centre of exchange; equally inevitably, when this became too cumbersome the cash was left behind, to be replaced by letters of credit as the investors became more confident in the stability of the Fairs and their administrators. So, during the thirteenth and early fourteenth centuries, trade boomed as water power and money combined to wind the economy higher and higher. Notarial registers of the times show that men, single women, widows, children in care, clergy, nobles – anybody with spare cash – were all investing money in some aspect of commerce, whose golden days must have looked as if they would never cease. They ceased soon enough, because in the very dynamism of the economy, in the clearing of the forests, in the ploughing of new fields and the founding of new markets, lay the seeds of its own destruction.

When times are good the population rises, and sooner or later, it rises too far. By the beginning of the 1300s some parts of Europe were already living on marginal land, producing enough to live on but not enough to lay aside for a rainy day. And that is just what it began to do:

rain. Early in the previous century what was later known as the 'Little Ice Age' set in. At the start it manifested itself as uncertain weather. Grain harvests began to fall short, and then to fail disastrously. There was widespread famine. All the time the weather got colder: winters became severe, summers cool. In this weakened state, Europe was ill-prepared to fight off an invasion – especially as the invader was almost invisible. It was a flea, carrying the bubonic plague, brought back from Caffa, a Genoese colony on the Black Sea. The Genoese there had become infected by rotting corpses tossed over the walls of the city when it was under siege by the Mongols. In 1347 a Genoese galley from Caffa was refused entry to Sicily, and eventually landed in Marseilles; by 1402 the plague had ravaged as far north as Iceland. Millions died – so many that there were not enough living to bury the dead. The plague brought everything to a halt. By the middle of the fifteenth century the population of Europe was half what it had been a hundred years before, and the common graves stretched for miles from every town and village. Artists of the period expressed the event in a new theme: the dance of death.

The inevitable consequence of the dereliction of the land and the general breakdown of supplies was that in some places inflation rose to 400 per cent in a generation. When the plague was over, however, everybody was better off in gross terms, since those who survived took what had belonged to those who had died. Real wages rose, labour was scarce and could command higher prices and better conditions. The new-found wealth filtered down to the bottom of society: by 1450 such things as the individual dish, the chair instead of the bench, a fireplace, even a dowry became common among the poorest families. The technology of production was still there, only now there were fewer people to operate it with more money to build it, so its true labour-saving qualities began to be exploited as production rose *per capita* by as much as 30 per cent. With the horror still within memory, Europe went on a spending spree in order to forget. The demand for luxuries rocketed, especially in modes of dress. A cursory glance at the paintings before and after the Black Death shows the difference. At all levels of society dress became so extravagant, and the power of the workers and their valuable labour so disturbing, that the kings began to pass sumptuary legislation to decree what each level of society might or might not wear. Still the spending went on. The huge expenditure on luxury cloth contributed to the general problem of the balance of payments at that time. In the case of silk, for example, increased consumption in France resulted in a massive outpouring of gold to Italy where the silk was produced, and it was in an attempt to prevent this that Louis XI set up his own silk-weaving industry in Lyons in 1466.

How the plague spreads. Rats infected with pasteurella pestis *bacteria are bitten by rat fleas which then carry the disease to other rats. At some point the flea bites and infects humans, who in turn infect other humans. The disease first shows as lumps (buboes, hence bubonic) in the groin or armpits. Vomiting and high temperature are followed by diarrhoea, pneumonia and death.*

The advancing wave of disease is shown (below) as the plague spread north through Europe in three years. Iceland was temporarily spared thanks to relatively rare contact with the Continent.

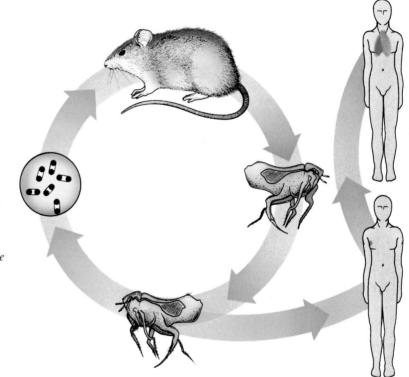

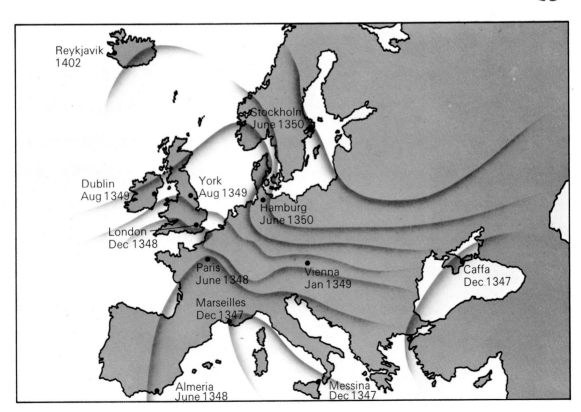

Reykjavik
1402

Stockholm
June 1350

Dublin
Aug 1349

York
Aug 1349

Hamburg
June 1350

London
Dec 1348

Paris
June 1348

Vienna
Jan 1349

Caffa
Dec 1347

Marseilles
Dec 1347

Almeria
June 1348

Messina
Dec 1347

While the upper classes bought silk, the peasants were able to afford linen, now in plentiful supply thanks to the earlier introduction of the horizontal loom and the spinning wheel. These two devices had considerably increased the production of both wool and linen. Linen was cheap to produce: to make linen thread all that was necessary was to harvest the flax, allow it to rot in water, dry it, beat the stalk to remove the fibrous material, and twist this together. The widespread and increasing use of linen in the late fourteenth century represents one of those crucial moments in the history of the process of change when the sequence of events suddenly changes direction and context. It caused one minor and one major alteration in society. As the linen was discarded when it wore out, a new social occupation was created. The bone collector, who had previously travelled from village to village collecting bones to be ground up for fertilizer, now included in his round the collection of linen rag, and became the rag-and-bone man familiar throughout the following centuries. Linen rag was, of course, excellent raw material for high-quality, durable paper.

At the end of the fourteenth century paper-making was a growth industry, functioning as it did in times of rising economic strength and consequent administrative paperwork. The paper-making techniques had originally come to Europe with the Arabs, who had picked them up when they overran Samarkand in A.D. 751 just after the Chinese had sent a team of paper-makers there to set up a factory. By 1050 the Byzantine Empire was importing Arab paper, and in Europe paper was first made in Muslim Spain at Xativa, south of Valencia. The first evidence of a paper mill working on water power is at Fabriano, in the Italian Marches, in 1280. The paper mills used the same basic technology as the fulling mills which preceded them. Rotten linen rags were pounded in water by trip hammers, together with gums, until a white pulp was produced. The pulp was laid in thin layers on wire mesh to drain, and then pressed in a screw press (similar to those used for pressing olives) until most of the moisture had been squeezed out of it, when it was hung to dry. The wire mesh itself is another example of the accidental way in which change comes about, because at the same time as the paper boom the sartorial excesses that followed the Black Death were giving employment to drawers of precious metal wire used in cloth-of-gold and -silver, so the wire mesh techniques were available for the paper industry to use, almost as if planned. Most of the early paper mills were set up in the foothills of mountains, to profit from abundant supplies of water power.

The demand for paper was high because it was comparatively cheap in relation to its competitor on the market, parchment. Between two and three hundred sheepskins or calfskins were needed to produce enough material for a large Bible, and the preparation of the skins was time-consuming and therefore costly. As the supply of the new linen rag paper increased, its price fell. By 1300 in Bologna, northern Italy, paper was only one-sixth of the price of parchment, and its price continued to fall. As Europe recovered from the plague and trade revived, the demand for manuscript went up to meet the increasing paperwork as the notaries produced the documentation that went with burgeoning business. The universities already had their own manuscript-copying departments, and in time private citizens went into the business. In the middle of the fifteenth century, for example, a certain Vespasiano da Bisticci ran a copying shop in Florence employing more than fifty scribes. Since the Black Death had killed off many of the literate members of the community, those who were left commanded astronomically high prices. The situation was clearly unacceptable: on the one hand scribes who cost too much, on the other, paper so cheap you could cover the walls with it. Craftsmen all over Europe must have been working on the solution to the problem, since in essence it was obvious: there had to be some form of automated writing.

It had been tried before. The Chinese had been using porcelain block-printing techniques to put characters on paper for more than a thousand years. In Korea, from about 1370 onwards, some form of interchangeable copper blocks had been used to do the same thing. In Europe the blocks were made of wood and used to print playing cards, calendars, prayers and occasionally capital letters in manuscripts. There were also a few books with entire pages printed by one block, but these were expensive and besides, the wooden blocks eventually wore down.

The credit for the great leap of imagination that followed is usually given to a goldsmith from Mainz in Germany called Johann Gansfleisch, better known to the world by his mother's family name which he adopted – Gutenberg. He first appears enrolled in the Strasbourg militia as a goldsmith member, just before his family returned to Mainz, which they had left during a rising against the members of the city's ruling class. Gutenberg's father had been an official in the bishop's mint in Mainz, and from him Johann had learned the tricks of handling soft metals, which were to prove vital when it came to printing. Apparently he had made an arrangement with three fellow-townsmen to make hand mirrors to sell to pilgrims leaving the city on a pilgrimage to Aix-la-Chapelle. Unfortunately somebody in the partnership had got the date of the pilgrimage wrong. After they had signed the agreement they discovered that the pilgrims would not be leaving until 1440, a year later than they had expected. Naturally there was some consternation, since presumably money had already been spent. At this point, Gutenberg promised to teach his partners 'a thing' he had been thinking about for some time. No more is heard of what this was, but later on, after another partnership with two other men called Fust and Shoeffer, the wrangling and lawsuits give the impression that Gutenberg's 'thing' had been a technique for printing. The main problem in developing automated printing was the creation of movable type, and in this Gutenberg had two great advantages. He knew how to work soft metals, and the language he used had only twenty-three letters in it (no j, v, or w). This may have been the reason the Chinese had never taken the matter of movable type further, since with a language made up of thousands of characters the problem was too immense to handle.

Gutenberg's aim was to make movable type that would not wear down easily, that would be uniform in size, and that would lie side by side accurately and uniformly enough to produce an even line of print. Each letter would have to be cast in an identical mould, and this was where knowledge of metal-working came in. The mould would have to be made of metal that would melt only at a much higher temperature than that from which the letter was to be made. Initially the mould was made in copper or brass, with the letter punched into it with a steel punch of the type used for placing hallmarks on precious metal. Early experiments may have been tried using a small box with the letter punched into the closed end, into which was poured the molten tin–lead alloy. The problem was that the box would have to be split apart to take out the cooled letter. Any doubts that have ever been expressed about Gutenberg as the inventor of printing are based on the fact that he appears to have solved the casting problem at a stroke, with a system that was working perfectly the first time we hear of it. Arguments run that he must have heard of other tests and designs, notably by men such as the Dutchman Laurens Coster, who were trying to do the same thing. Whoever designed the final mould, it is a work of genius. In order to save breaking the mould every time a letter was cast, a mould was made in three sections that slid together and were held in place by a large curved iron spring; after casting these parts slid apart, to be used again. The product was called mitred type. The greatest advantage was that the mould made all letters of the same dimensions. The space taken by a cross-section of each letter was identical, as were the space types, so that the printing was regular. The form – the stalk behind the letter face itself – was the same height, so that each letter would press on the paper with equal firmness. The problem of uniform pressure was minimized by the fact that the paper used was thick and soft, and was wetted slightly before printing, and it was further helped by the use of a linen or grape screw-press to push the letters in their bed-shaped box down on to the paper with equal pressure all the way across. But it was the interchangeability of the letters that lay at the heart of the new invention.

How uniform-sized movable type was first made. Left to right, the steel letter punch was prepared, then used to hammer the impression of the letter into a copper matrix. The lead–tin–antimony alloy was poured into the mould, structured so as to produce the typeface raised on a shoulder and stalk of uniform height.

In the traditional press shown below, the paper was placed in the rectangular holder (right), a protective frame folded over and above it, and the paper was then screw-pressed on to the prepared box of inked type.

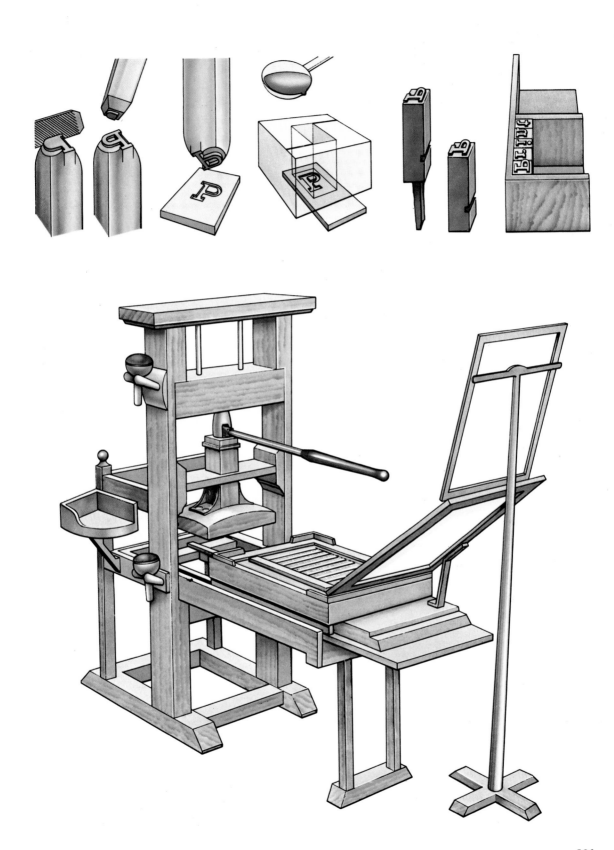

The earliest dated example of the new technology in action is the Mainz Psalter, printed at the order of the Archbishop in 1457. (There are other pieces of early printed material, but they are undated.) The introduction displays the pride felt by the printers at their achievement. It reads: 'This volume of the Psalms, adorned with a magnificence of capital letters and clearly divided by rubrics, has been fashioned by a mechanical process of printing and producing characters, without use of a pen, and it was laboriously completed, for God's Holiness, by Joachim Fust, citizen of Mainz, and Peter Schoeffer of Gersheim, on Assumption Eve in the year of Our Lord, 1457.' If you look closely at the original text, you will see that there are many more than the basic twenty-three letters. There are lines above some letters. Others are joined together, and others abbreviated in forms that are no longer recognizable. This is because the first printers had to produce the nearest thing to a manuscript they could, or the work would not be accepted by readers used to the script techniques of the scribes, which included abbreviations and even a form of shorthand. To have printed the texts in full would have insulted the reader's intelligence, as a printer would today if he printed 'To-day Ma-ry is go-ing to vi-sit her un-cle'.

The advent of printing, whether due to a German or a Dutchman – or even, as has been suggested, to an Englishman – was one of the most critical events in the history of mankind. Printing first and foremost made it easy to transmit information without personal contact, and in this sense it revolutionized the spread of knowledge, and craft technique in particular. 'How to do it' books were among the first off the press, written about almost every field of human activity from metallurgy, to botany, to linguistics, to good manners. Printing also made texts consistent, by ending the copying errors with which manuscripts were rife. In doing so it placed on the author the responsibility for accuracy and definitive statement, since many more people were now likely to read his material who might know at least as much about it as he did himself. This in turn encouraged agreement on the material, and because of this, spurred academic investigation of subjects and the development of agreed disciplines. Just as learning became standardized, so did spelling. Authorship became an object of recognition, and this led to the concept of 'mastership' in a subject, which in turn led to the fragmentation of knowledge into specialized areas, emphasizing the separation of the 'expert' from the rest of the community. The earliest books would have been read by men who could doubtless as easily have turned their hand to the lyre or the sword or the pen or the architect's drawing, and it may be said that with the coming of the book they were the last generation to be able to do so. The new texts also conferred prestige on the inventor, who could now publicly claim association with his invention and expect to

Part of the Mainz Psalter, showing the manuscript abbreviations carried over into early printing – for example, in line two, the line above 'nris' makes the word 'nostris'.

A sixteenth-century printer's shop. Left, the compositors make up the page. Paper arrives at the door, behind a man inking the plates. Printed sheets hang drying above the press operator (right). The bespectacled proof-reader stands behind a compositor, checking for errors.

be identified with it. And as the books began to circulate, carrying ideas to readers who no longer had to have access to a manuscript copyist producing rare and expensive editions, the speed of change born of the interaction of ideas accelerated.

The coming of the book must have seemed as if it would turn the world upside down in the way it spread and, above all, democratized knowledge. Provided you could pay and read, what was on the shelves in the new bookshops was yours for the taking. The speed with which printing presses and their operators fanned out across Europe is extraordinary. From the single Mainz press of 1457, it took only twenty-three years to establish presses in 110 towns: 50 in Italy, 30 in Germany, 9 in France, 8 in Spain, 8 in Holland, 4 in England, and so on. By 1482 the printing capital of the world was Venice, and the busiest printer there was a man called Aldus Manutius who used to have a sign outside his shop saying 'If you would speak to Aldus, hurry – time presses'. He had good reason. No single printer did more to spread the printed word than he. Aldus knew that his market, and the market of all printers, lay not in the production of expensive, commissioned editions of the Bible or the Psalms, but in an inexpensive format that could easily be carried in a man's saddlebag wherever he went. So Aldus made his books small, and cheap. The Aldine Editions, as his new format was called, were the world's first pocket books, and they sold faster than he could produce them. Nearly half his workers were Greeks, exiles or refugees from the Byzantine Empire after the fall of Constantinople to the Turks. So it was that with the help of his translator-craftsmen, Aldus began the task of translating the Greek classics. When he died, in 1515, no major known Greek authors remained to be translated. Whatever happened in the Greek world, Aldus had ensured that the classical authors would not once again be lost to the West, as they had after the fall of Rome.

As Europe recovered from the Black Death and the towns began to grow and prosper again, the close proximity of Italy to Greece had led, by the beginning of the fifteenth century, to a quickening of interest in Greek scientific thought and philosophy. Many exiles from Constantinople had come to teach in Florence and Venice. Among other things they taught of the work of Hero of Alexander in the second century B.C. on pneumatics and hydraulics. To a nation interested in renewing water supplies to its cities, and spending its new wealth adorning those cities, the Greek engineers and architects provided a stimulus for a revival in things classical. This is the revival known as the Renaissance. Aldus had ensured that the texts were readily available. Suddenly, through printing, the past became something tangible rather than a myth passed on by word of mouth, and in the writings of the Greeks the Italians found a civilization to emulate. In one sense, with the attraction of the Renaissance reader to Greek ideas, and the Greek separation of intellectual from manual work, the Renaissance may be seen as a period of technological slow-down compared to the effervescence of the Middle Ages. Interest on the part of the patrons switched from practice to theory. Hero's pneumatic toys – temple doors that would automatically open, birds that would sing on steam – may have displayed certain natural laws in action, but the princes of the sixteenth century, most of them in politically precarious positions, were more concerned with the use of such clever devices to enhance their public prestige. Their interests were well served by engineers and metal-workers, for whom the court offered the lure of social advancement.

The craze for automata that followed had its origins in the systems described by Hero for using water and air pressure to drive machinery. As has been said, there was a considerable amount of land to be re-claimed in northern Italy, and it was in Italy that the fun-loving aristocrats took things a stage further with the complex and expensive water gardens. The best known of these, and the one that served as a guide to the many that followed it, was the water garden of the Villa d'Este, built in 1550 at Tivoli, outside Rome, for the son of Lucrezia Borgia. The slope of a hill was used to supply fountains and dozens of grottoes where water-powered figures moved and played and spouted. Within fifty years of the Roman example there were water gardens all over Europe, and princes and bishops would delight in switching secret valves to deluge their guests with soot or spray them with water. One of the finest of these gardens still exists today at the Château Merveilleux of Hellbrun, outside Salzburg. It is full of performing figures of men and women and animals, set in corners of the garden where fountains turn on and off unexpectedly, or operating in the intricate and quite amazing theatre of puppets, run by water power.

One of the dozens of fountain systems at the Villa d'Este, Tivoli, where the height of the jets was altered by valves and sluicegates controlling the pressure of the water supply coming down the slope above. Inset: The water garden at Hellbrun: (above) the Orpheus grotto, and (below) the mechanical theatre. The tiny figures, including guards march-ing up and down, a butcher killing a calf, and the town band playing, are all operated by levers and rods connected to cams on water-powered shafts.

Automatic music was produced in the same way in water organs, which became popular enough for Mozart and Haydn to compose music to be played on them. Air pressure in a tube was produced either by water-operated bellows, or by the pressure of water at one end of the tube. The air was then released into organ pipes by valves opened selectively according to a pattern of cams rotating on a cylinder. The music merely had to be translated into the pattern to which the cams, or pegs, were inserted into the side of the cylinder. Since the cylinder had rows of holes for the pegs, the same cylinder could be re-pegged to produce different tunes, exactly as had the Mechelen carillon cylinder. It was the automated organ, widely in vogue at the end of the seventeenth century, that was to solve a major problem for the French silk industry in Lyons, set up by Louis XI in 1466. At some time in the fifteenth century a silk-worker called John of Calabria, an Italian immigrant, had introduced a new kind of silk-weaving loom, which made the process somewhat easier. The basic difference between weaving wool (or linen) and silk lies in the fineness of the latter's threads. Because there are so many of them to the square inch, and the material itself was already expensive, silk was costly. By the time the automated organ was in use, John's drawloom, as it was called, was already working in Lyons, with a few modifications. First a pattern was drawn on squared paper, to correspond to the pattern desired on the finished weave. Squared paper was used to show which

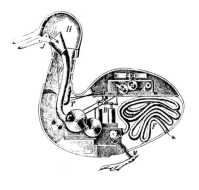

Vaucanson's automated duck. It quacked, ate, drank, and excreted.

An early seventeenth-century automated organ. The water wheel turned a pegged cylinder; each peg moved a lever which opened the top of an organ pipe, releasing air compressed by the weight of water in the reservoir – and the organ played.

Right: A mid-nineteenth-century shawl in the now familiar Jacquard pattern. These designs were inspired by those on cotton imported from India and, prior to the Jacquard loom, were too intricate to mass-produce economically.

Below: A 'bizarre' pattern in silk. At the beginning of the eighteenth century these complex designs, inspired by silks from the Far East, required immense care in weaving. Drawboys' mistakes were ruinously expensive.

of the warp threads had to be lifted each time the weft thread was passed between them. In this way, all the warp threads to be lifted at the same time could be attached to a common cord. When the cord was pulled, all the threads attached to it would lift in unison. The job of pulling these cords was left to children, commonly called drawboys, who worked long hours and often became tired enough to pull the wrong cords, with disastrous results. The weaver would only know of the mistake well after it had been made, when it was too late to correct it.

By 1725, as the desire for new designs grew to proportions equalled only today, some way had to be found to prevent such costly errors. Lyons had by this time a reputation to live up to: that of mistress of fashion. It was a reputation that the French made a monster. A contemporary painter who was commissioned to do a series of oils of people of different nationalities in their country's costume painted the Frenchman naked, with a pair of scissors in his hand, waiting to see what tomorrow's fashion would dictate before committing himself to wearing anything. While the master weavers of Lyons might just have been able to keep up with the changing fashion in dress design, they could not do this and produce all the other materials required, such as hangings, tablecloths, bedcovers and so on. The answer to the problem came from the son of an organ maker, Basile Bouchon.

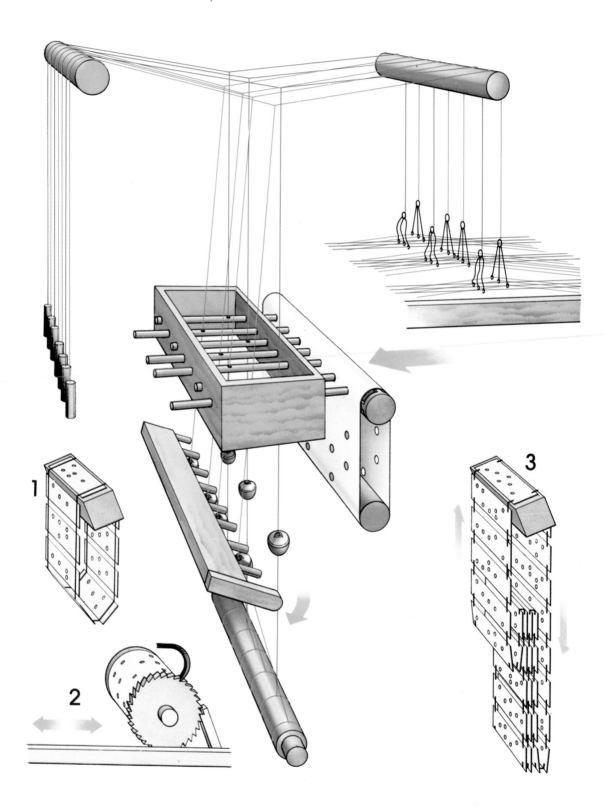

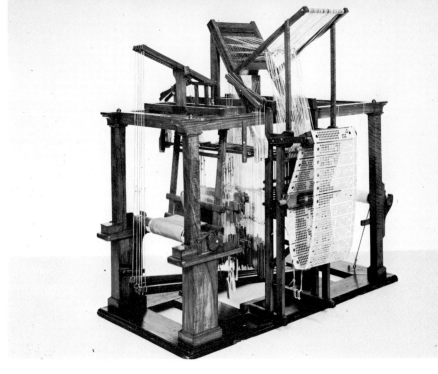

Right: A 3-foot high model of Falcon's loom, showing how his division of different stages of the pattern on to separate cards made it easier for the weaver not to make a mistake in positioning the matrix of holes.

Left: The various stages in the development of the automated loom. The main drawing shows Bouchon's system. When the paper roll is pressed against the horizontal rods, a hole in the paper admits the rod, leaving it unmoved. Where there is no hole, the paper pushes the rod which carries the wire threaded through it laterally to the left, causing the ball on the wire to engage under the teeth of the comb at the bottom. When this comb is rotated downwards it pulls on the engaged wires, which in turn pull the groups of thread to which they are attached. These threads lift, permitting the shuttle to pass underneath.
1. Falcon's improvement on the paper roll.
2. Vaucanson's cylinder.
3. Jacquard's 'endless' chain of cards, and prism.

Bouchon must have known of the early stages in the construction of organs, which involved placing pegs on a cylinder according to a series of holes cut in a piece of squared paper which was fitted round the outside of the cylinder to act as a pattern. This was the concept that he applied to the silk loom.

Bouchon's idea was improved upon in 1741 by one of the finest automata makers in France, Jacques de Vaucanson, who returned to the use of a cylinder. Into this cylinder he punched rows of holes, and then placed the paper with its holes in it round the cylinder, so that the wires to be left unpushed would slide into the holes in the cylinder. The point of this was that the cylinder was set in a frame with a ratchet, and each time it moved against the wires, it would click on one row to the next set of pattern holes in the paper. Naturally enough the silk weavers of Lyons saw in the new automated looms a threat to their livelihood, and there were riots. Vaucanson's loom lay unnoticed in the Paris Museum of Arts and Crafts for over fifty years, until, in 1800, a silk weaver called Joseph Marie Jacquard was asked to put it together again. In doing so he made a few minor adjustments. He went back to an earlier idea and replaced the paper with cards, each one carrying a separate section of pattern. The cards were mounted on a belt carrying each one to a point where it wound over a prism, shaped rather like a squared-off cylinder, which revolved, as Vaucanson's cylinder had done, with a ratchet system. All the weaver had to do to extend the pattern was add more cards. For this Jacquard got all the glory, and the loom to this day is known as the Jacquard loom. All over Europe the workers' reaction was the same: they smashed it. Eventually the loom was taken up and used extensively, but it took many years to gain acceptance.

The idea of using paper with holes in it spread to engineers, and in 1847 Richard Roberts in England adapted the idea to control rivetting machines working on the construction of the Menai Straits iron bridge in Wales. The machines used multiple rivetting spindles, a selection of which would be brought into action at any one time, depending on the shape of the plate to be rivetted. The same machine was used in the building of the great iron ships that sailed the Atlantic, taking huge numbers of immigrants to America. The increase in the population of the United States was such that in 1880 it took officials carrying out the 1880 census eight years to complete the counting, and the flood of people still pouring into the country meant that things could only get worse for the census-takers. The man in charge of the health statistics division of the census was a lieutenant-colonel in the U.S. Army, attached to the Surgeon General's office. He was a doctor called John Shaw Billings, and while watching the early returns being counted in the census building in Baltimore in 1880, he suggested to a young engineer attached to his division that there ought to be a more auto-mated way of doing the counting. His idea was to use Jacquard's cards to mechanize the filing of the immense quantities of data being collected. Herman Hollerith, the engineer, went away and thought about it, and then returned to ask Billings if he wished to be involved in working it out. Billings replied that all he cared about was solving the problem, and that Hollerith could go ahead alone.

Hollerith took cards the size of dollar bills – because there were already dollar-bill holders on the market, so there was no need to design one – and punched holes in the cards in predetermined positions relating to the type of data being recorded: sex, age, size of family, location, date of birth, nationality, etc. The number of separate bits of data thus recorded on the card was extremely large. When the 1890

Steerage passengers on board an emigrant ship to America in 1872. Travellers brought their own cutlery, crockery and mattresses. Conditions varied from bad to appalling.

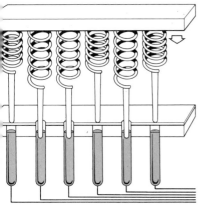

census was taken, the machines Hollerith had designed to go with his cards were used for the first time. The census was completed at twice the speed of the previous one, at an estimated saving of half a million 1890 dollars.

Hollerith's (or Jacquard's, or Vaucanson's, or Bouchon's) punch-card memory was, within a few years of the beginning of this century, being used in tabulators and calculators, and finally in electronic enumerating machines. Today the data goes in and comes out via keyboards and video display units, but the concept is essentially the same. Bouchon, and the automata makers before him, had hit on the binary code, the 'yes or no' operating code used by the computers that run the modern world, and without which that world would grind to a halt.

The genesis of the modern computer, Hollerith's tabulator. Recording the information stored on cards involved an extension of Bouchon's loom card idea. A matrix of electrified sprung wires set in the punch being used on the right was pressed on to the card. Where there was a hole (see diagram above) the wire went through, making contact with the mercury in a cup below. The signal then activated the relevant counter on the clock-faces (right), causing the hand to click round one space. Each cabinet would count a single characteristic, first as it appeared on each State's dial, and secondly in an overall total on the single dial representing the entire country.

114

5

The Wheel of Fortune

The great Arab astronomers, who brought a sophisticated mixture of Greek and Indian science to the medieval West. To them we owe our system of numbers and the use of the decimal in calculation. Here the observers are using, among other things, the astrolobe, which was perfected by Arab instrument-makers.

It was between the two world wars that the computer progressed from being a machine that could add faster than a man to the stage where it began to hold the future in its entrails – and it did so because of aeroplanes. Until then the fastest thing an artillery officer had to hit with his gun battery was probably an armoured car. The aeroplane presented him with the problem of aiming off, to compensate for its speed – in other words, he had to aim at where the plane would be when the shell arrived. Doing this accurately involved knowledge of such things as the speed of the plane, the weight of the shell, the power of the gun, the wind speed and direction, the temperature of the air throughout the shell's trajectory, and so on. In the 1940s the U.S. Army produced a series of artillery tables that offered artillery officers every possible mix of conditions, and told them where to point their guns. Computers were used to produce these tables, and in doing so they performed their first predictive function, computing the destruction of a target which, at the time the tables are used, is not yet there.

It is in its predictive capacity that the computer has had, and will continue to have, its most profound effects on the modern world. It can do other things: it operates as an immensely fast calculator, without which much modern business could not function as it does: it saves gigantic amounts of paperwork and the labour of thousands of people who would otherwise be needed for it; it can, through being programmed with a memory, direct other machines. But what will the existence of the computer mean for us, as it is increasingly able to show us the future?

The computer works on a 'yes or no' basis. In electrical terms, this becomes the presence or absence of a signal. On is 'yes' and off is 'no'. This binary (twofold) form of information can be used to handle large amounts of data using relatively few signals. For instance, five on/off signals can be used to represent the alphabet. Off off off off on is 'A'. Off off off on off is 'B'. Off off on off off is 'C', etc. By first moving the 'on' signal up the group of signals then introducing two 'on' signals which are moved, and so on, a group of five signals can provide $2 \times 2 \times 2 \times 2 \times 2$ ($= 32$) different ways of arranging the signals. So, a five-signal group can handle 32 bits of information. And since, to all intents and purposes, on/off electrical signals can be 'read' at instantaneous speed, the computer can handle billions of signals in a second.

The two factors that make the computer worth using are the speed with which electricity flows, and the immense amounts of data that can be handled using the binary code. For this reason the computer can be used to solve problems whose calculation would take a human longer than a lifetime. Turn the problem into a mathematical equation, and the computer will solve it or say why it cannot be solved. And all problems can be expressed in terms of yes/no elements. There is no need, for instance, to pour one chemical into another in order to see what will happen. Tell the computer the rules of behaviour of all chemicals and it will mix them for you, giving the results on paper or on a visual display unit, using the same binary code to display letters as to mix chemicals.

This kind of ability is particularly valuable in predicting hazardous situations, when the computer will print 'Bang!' without anybody having to get blown up. It is also useful in working out such problems as the balance of profit and loss relating to the choice of site for a major corporation, or new city, taking into account variables in the availability of resources, transportation requirements, market availability, cost of construction and so on, and essentially printing out 'yes' or 'no'. Virtually every community decision is made like this. It is only in the face of obdurate ideological or emotional pressure that the computer is likely to go unused in the future, which is why concern is at present expressed regarding the extent to which the computer is being used to organize the community. Because of the increasing number of human activities that involve the use of computers, more and more data are being compiled in the databases that relate to aspects of the individual's life that he once regarded as inviolate and private.

The virtue of the modern world is creditworthiness, and to have it individuals must tell more of themselves than they may wish: their bank balance, their marital state, their criminal record, their character as assessed by other members of the community, and so on. And as the credit card, or 'electronic fund transfer' system, spreads – as it will – how will we live without credit in a world that will not accept cash? The concern is not so much that databases may or may not hold vast amounts of private information about us, but that such data may be predictively used by the computer on a mass scale. For a system handling enormously complex equations in trillionths of a second, the *matter* of the equations is of no consequence. The computer will mix races as efficiently as it will mix chemicals, and as easily show what must be done to avoid the bang, in either case. If detergent preferences can be changed by a television commercial, why not political attitudes? The computer holds the future within it, and foreknowledge is power. The question is in whose hands does this power already lie?

The last time Western man became enmeshed in a system that would tell the future, and confer power on those who could operate it, was in A.D. 765. The system was astrology, and power lay in the hands of those who could foretell eclipses and the time of floods. Astrologers had read the sky since early Sumerian times, in the twenty-third century B.C., but matters came to a head in A.D. 765, in the newly founded city of Baghdad, when the second 'Abbasid Caliph, Al Mansur, became ill with a stomach complaint. The Caliph's predecessor had decided to build the new capital with circular walls on a healthy site, which by chance was about 150 miles from a monastery in the mountains of southern Persia, at a place called Jundi Shapur. In the monastery was a famous hospital and medical school, and it was to this that the Caliph sent for help. The head of the school, a Christian monk called Jirjis Bukhtu Yishu, came down to Baghdad and cured Al Mansur. But it was how he cured the Caliph, and what else was going on in the monastery of Jundi Shapur besides medicine, that was to have such a significant effect on the history of Western Europe. By the time of Al Mansur the monks had already been at Jundi Shapur for over a hundred and fifty years. Previously they had been at Edessa, further north, until the Byzantine Emperor had closed the place and moved them on. And they had been in Edessa since A.D. 431, when they had moved there from Constantinople with their leader, Nestorius, the ex-patriach of Constantinople who had been exiled for heresy. At Jundi Shapur they ran a medical establishment, and translated classical Greek and Persian texts into their language: Syriac. The texts they translated included much of the Hellenistic scientific work that had been written and taught at the great university in Alexandria, which included astronomy, medicine, mathematics, philosophy and astrology.

When Al Mansur was cured he persuaded Bukhtu Yishu and some of the monks to come to Baghdad and set up a hospital. He and his descendants were to run the hospital for more than two hundred years, exerting a considerable influence on the course of Arab medicine, and, through their translation, on Arab philosophy and science. The Arabs were traditionally interested in astronomy, since for religious reasons it was necessary to know in which direction the holy city of Mecca lay, and at exactly what time the muezzin should ascend the minaret to call the faithful to prayer. So Al Mansur and his successors encouraged the translation of the Syriac versions of the Greek and Persian star tables – compilations of star positions at all times of the day and night throughout the year – into Arabic. Tradition has it that there was another arrival at the Caliph's court: a man who claimed to be able to calculate the movement of the stars and predict eclipses. He had come from India, from the observatory at Pataliputra in northern India where, from the fourth century onwards, the Hellenistic science of astronomy had flourished in the wake of the empire of Alexander the Great, who had taken much of Greek culture there. Meanwhile the Indian astronomers, encouraged by the Gupta dynasty, had built up a record of observational data called the *Siddhartha*, the manual of the stars. It was this manual which the new arrival in Baghdad translated into Arabic. The introduction of Indian mathematical techniques into the Greek tables brought the decimal system to the science.

It is easy to see how the astrological aspect of this knowledge of stellar behaviour should have wielded such a powerful influence in ancient times, and even until as late as the seventeenth century. Primitive peoples who observe over decades that the position of the sun in the sky relates to the changing seasons, will soon ascribe the fertility of the soil to powers in the sky. As data on the sky increased, astronomer-priests were able to foretell the times of one of the most awe-inspiring events in the heavens, the eclipse of the moon, and by Alexander's day they could plot the movement of the planets and stars to within minutes of arc and time. There are in the sky twelve clusters of stars ('con-stellations'), strung out more or less equidistant along the plane of the ecliptic – the path that the sun appears to follow – so the ancient astronomers had divided the sky into twelve sections. These sections became the basis for the zodiac, and for the twelve hour night and day. Early on, the shape of these constellations led to their assuming animal or human identities: fish, archer, ram, lion and so on.

An early Christian adaptation of the zodiacal system. The principle of predestination inherent in astrology posed a problem for those who believed in God-given free will: they solved it by placing Christ at the centre of the scheme.

As the pseudo-science of astrology became more complex, the powers of the shapes in the sky became related to objects on earth: plants, animals, minerals, the weather, and every aspect of the condition of man. As the ancients observed the magic perfection of the heavens whose movements repeated themselves with ineluctable certainty year after year, it appeared obvious that the vagaries of human life too should be subject to such a plan.

Each of the twelve sections of the sky was therefore given power over some aspect of life. The first 'house' (as each section of the sky was called), that on the eastern horizon, was the house of life. The second house, next up in the sky above the first, was the house of riches; the third, fraternity; the fourth, relationships; the fifth, children; the sixth, health; the seventh, marriage; the eighth, death; the ninth, religion; the tenth, dignity; the eleventh, friendship; the twelfth, animosity. As the Earth turned, it appeared that the planets and the constellations moved through each of these twelve houses in turn, and since each star or planet had significance for almost anything the astrologers could dream up, its influence on a man would depend on whatever house it was passing through at certain critical moments in his life. Thus Jupiter (whose astrological sign is still placed at the beginning of a doctor's prescription!) meant 'male', 'white', 'land' and 'electrum' (a mixture of gold and silver), and other things as well. An astrologer who knew the exact time of a child's conception could tell its expectant parents that, for example, if it had been conceived while Jupiter was in the second house (riches) the baby would be a successful, fair-complexioned, landowning male jeweller. It was on the basis of such belief in predictive abilities that Herod asked what time the star appeared over Bethlehem.

These predictive abilities relating to the human condition were also used by physicians in their diagnoses and cures, since disease and treatment were closely linked with the heavenly influences at work at the onset of the complaint. Blood was let, for instance, only on Tuesdays and Wednesdays, since these days belong to Mars and Mercury which both had differing influences on the blood. The Sun influenced the right eye; Saturn, healing; Mars, blood; Mercury, blood, the tongue, gullet and speech organs. Wherever the pain might appear, the planets told the doctors where the real seat of the trouble was. Treatment was equally dictated by the sky, since every planet and star had its herbs and metals. By the time Al Mansur became sick, this medical side of astrology was sophisticated enough to have been used to cure him. It demanded an exact knowledge of celestial mechanics, and the Arabs, with their fanatic concern for health and hygiene, avidly sought out any manuscript that would enhance that knowledge. One of the texts translated was Ptolemy's System of Mathematics,

Zodiacal Man, showing the parts of the body over which the specific signs had particular power.

written in the second century A.D. This work, the greatest collection of astronomical data known, was dubbed by the Arabs *Almagest*, 'the great work' – the name by which it is known today. Apart from being a catalogue of stellar movement, the work also gave details for the construction of instruments for making exact observations of star position.

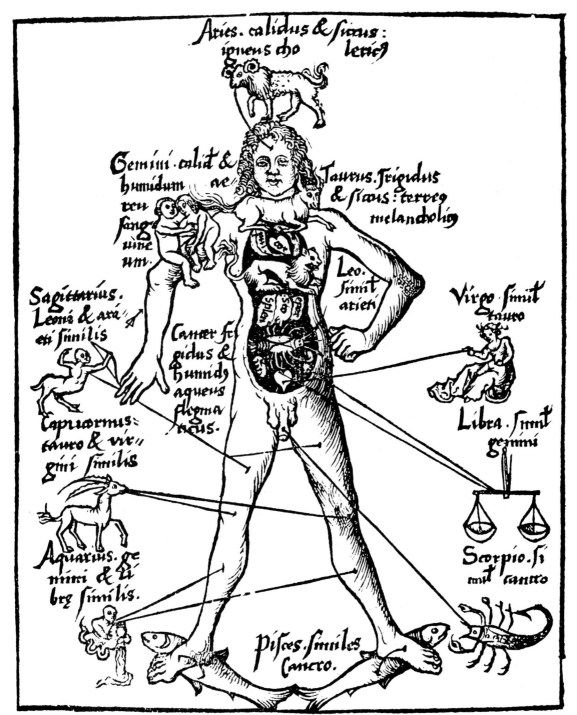

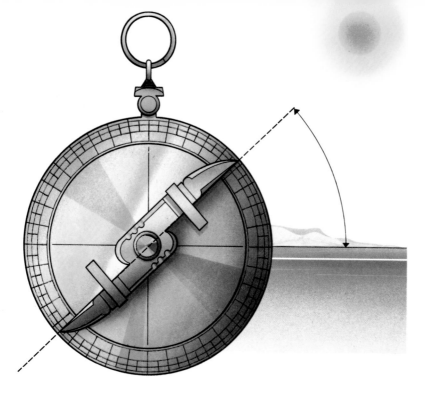

By the ninth century the Arabs had perfected the astrolabe, the instrument which was to be the basic astronomer's tool for the next seven hundred years. In all probability they inherited it directly from the astronomical texts they were translating. It was most often made of bronze, and was essentially a replica of the stars in the sky, which could be moved to show where the stars would be at any time in the year. On the reverse side of the circular instrument was a set of gear wheels which could be used to place the sun and moon in any position. By setting the stars, or the moon, or sun, in the position the user of the astrolabe saw them, he could then read off the date. The astrolabe performed one other vital function. It could be used to tell position, since, if used in conjunction with the star tables, it would tell the observer where he would have to be to see such-and-such a star on a certain date in that position in the sky. As the Arabs moved west, across North Africa and into Spain, they took their knowledge with them. At the great library in Cordoba in the tenth century there were over 600,000 manuscripts, more than the total number in France at the same period. News of this vast fund of information spread to Europe, and as the continent recovered from the anarchy and confusion of the previous five hundred years, European scholars began to take an interest in astronomy. The first appears to have been Gerbert, later head of the cathedral school at Reims, and later still, Pope. This low-born priest was a brilliant student and in 940, at the age of seventeen, he had been sent to northern Spain, still under Christian control, to study with Otto, Bishop of Vich in Catalonia. It may have been during this time that he sensed just how extensive was the corpus of Arab

The astrolabe in its simplest form, used to measure the altitude of the sun, moon or stars. The pivoted limb has sights at each end. The instrument was suspended by the ring at the top, and the limb moved until the beam of sunlight entering the hole in the upper sight coincided with the hole in the lower sight. The angle at which the limb stood when this happened – that of the sun's altitude above the horizon – was read off the graduated scale around the circumference. At night a sighting was made directly through the holes.

knowledge. Whether or not he actually went to Cordoba is not known, but about 985, back in France, he wrote to the monk Lupitus in the county of Barcelona to ask for 'the book on astrology'. Following this he wrote a treatise on the astrolabe, and himself built a planetarium, a globe of the night sky, and a wooden cap incorporating a sighting device for finding stars. Gerbert was the first European to teach and argue for 'the influence of the stars'. He was not long alone. In the early twelfth century an Englishman, Adelard of Bath, and a German, Herman of Carinthia, both made trips to Muslim-held countries and brought back translations of Arab science and philosophy.

But the event that must have done more for the intellectual and scientific revival of Europe was the fall of Toledo in Spain to the Christians, in 1105. The Spanish libraries were opened, revealing a store of classics and Arab works that staggered the Christian Europeans. In 1130 the Archbishop, Raimund, opened a school of translators which concentrated on the work of the Arab philosopher and scientist Avicenna, thus bringing the logic of Aristotle, lost to the West for centuries, back into the universities of France, Italy and England. No man did more to astonish Europe with the new-found knowledge than Gerard of Cremona, in Italy. In the middle of the twelfth century he went to Toledo 'looking for Ptolemy', and there he found him. When Gerard died, in 1187, he had translated the *Almagest* into Latin, as well as more than ninety other texts. The *Almagest* took Europe by storm. In 1276 the King of Castile and León, Alfonso the Wise, ordered a new edition of the star tables, updated from Ptolemy's and those of his later Arab commentators. The Alfonsine Tables were to become standard astronomical reference works for three centuries. They immediately spurred interest in astronomy. By 1281 tables had been worked out with their reference position at Mechelen. In 1292 tables were written for Paris, and in 1300 for Montpellier.

King Alfonso and his school of translators at work in Toledo, from a contemporary manuscript. His principal assistants were Jewish scholars. Apart from astronomical works, the major work produced by the school was a treatise on precious stones.

The surge of interest in astronomy was matched by developments in the introduction of Arab medicine into southern Europe, at the School of Salerno, near Naples. Already established by the time Gerbert was studying in Spain, by A.D. 1000 the School was known throughout Europe. The basic medical curriculum came from the Arabs, thanks to a renegade Muslim turned Christian called Constantine the African, so named from Tunisia, his country of origin. After travelling and studying in Baghdad, India and Constantinople he returned to his native city of Carthage, leaving again after some trouble with the local medical fraternity. From there he went to Salerno, to the court of the new Norman ruler, Robert Guiscard, and spent some time there as secretary to the count. It seems odd, today, that a man should simply arrive on a ship and take up a high administrative post in a country whose language he does not speak. But southern Italy at the time was a melting-pot of race and tongue. Half of it was still Greek-speaking, under the direct rule of Constantinople. (Indeed, it is still possible today to find isolated villages where the local dialect bears traces of Greek influence.) In Sicily, where Robert's relatives were the rulers, the local language was Arabic, as well as the Sicilian form of Italian. And everywhere, anyone with pretensions to literacy spoke Latin. There was ample room for any newcomer with knowledge to peddle.

Thirteenth-century Arab physicians hold a clinical consultation over a patient. Most of their medical knowledge had come from translations of earlier Greek texts. It was often the practice to carry small books of diagnostic information hung on a belt round the doctor's waist.

Medical practice in the early thirteenth century, illustrated in a manuscript from the School of Salerno. Above, treating an ear infection. Above right, one doctor encourages his patient to inhale fumes from boiling herbs and ointments, while (far right) his colleague examines a severe case of skin disease. The Salerno physicians were drawn from all over Europe and the Mediterranean; one of them was a woman.

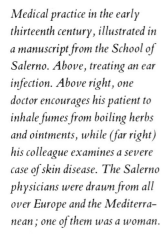

After a short stay in Salerno Constantine moved to the Abbey at Montecassino, where, together with another Arab, John the Saracen, he translated the Arab medical text called the *Royal Book*, written in the tenth century by Ali Ibn al-Abbas, physician to the Caliph in Baghdad. The translation into Latin was called the *Liber Pantegni*, and contained twenty chapters on subjects including urinalysis, dietetics, gynaecology, surgery and the capillary system. Constantine himself became something of an expert in urinalysis and was known as 'the man to whom women come with urine that he may tell them what is the cause of the disease'. By the twelfth century the School at Salerno was able to use this type of diagnosis to identify no less than nineteen different diseases. However, it was in the field of surgery that the School was to gain its reputation when, with the first Crusades in the eleventh century, it became a transit hospital for Crusaders too ill to complete their journeys home. When the Normans, under the leadership of Robert Guiscard, had consolidated their position in the south, Salerno became effectively the first university in Europe. The medical course consisted of three years' general study, and then four more on medicine, with a final eighth year when the student worked as an intern under the supervision of a physician. Specialization in surgery demanded an extra year. In the twelfth century the School formalized all it knew in the publication of a massive set of medical precepts, called the *Rule of Health*.

The Rule became the basis for all medical teaching in Europe for the next four hundred years. By the time it was superseded, in the sixteenth century, it had gone through 140 editions and been translated into all European languages. (Oddly enough the Rule is dedicated to a king of England who was never king, Robert I. Having spent some time at the hospital in Salerno recovering from a wound in the arm, he returned to England three weeks after his brother had taken the throne and crowned himself Henry I.) The Rule covered hygiene, nutrition, medical remedies, anatomy, pathology, therapy and pharmacology. It consisted of 2500 verses, written thus for easier

126

memorizing. The rules on diet and hygiene are remarkably modern: 'Get up at dawn and wash at least the hands and the eyes, preferably in cold water. Then comb the hair and brush the teeth. . . .Eat a hearty lunch but a light supper. . . . Try to get never less than six hours sleep. . . . Never miss an opportunity to urinate.' The main effect that the Rule had on European thinking was through its instruction in the 'humoral' theory of disease. This had first been propounded by the Hellenistic writer Galen, and then taken up by the Arabs. The theory stated that the body had four basic humours: blood, phlegm, yellow bile and black bile. These humours were associated with the material substances in the world around: blood was associated with heat, phlegm with cold, yellow bile with dry and black bile with wet. Fire was hot, and so was summer. Water was cold, and so was winter. Air and spring were dry, earth and autumn were wet. The connection with astrology was close enough for much of the common-sense medical knowledge in the Rule to give way to the mumbo-jumbo of the humoral theory of treatment. Nonetheless, by the mid-twelfth century Salerno was able to handle fractures, wounds, fistula, stones, hernias, ulcers, abscesses, skin disease, urine problems and general surgery; anaesthetic, in the form of a sponge soaked with the juice of a narcotic plant, was used regularly.

As the new medical knowledge spread into Christian Europe from the south, new astronomical knowledge came in from the west, from Spain. Suddenly the Europeans had tools with which to examine the universe, and tools to investigate the workings of the human body. Small wonder that the two principal disciplines in the new universities of Paris, Oxford, Cambridge and Montpellier were medicine and astrology. The Arab medicine worked better than whatever primitive techniques the Europeans had used before, and the revelation of the mechanical universe acted as a stimulant to philosophy.

Thanks to the efforts of Charlemagne in the ninth century, by 1050 every cathedral in Europe had a school, and as more and more of the Greek classics were translated their teaching became more and more speculative. The spread of the schools, the revival of learning, the growth of a money economy as trade and commerce spread, and above all the arrival of the new sciences brought a renaissance in Europe which expressed itself in every sphere of human activity. It must have seemed like the dawn of a new age, with nothing ahead but higher standards of living, wider contact with the rest of the world, and peace and security after centuries of war and turmoil. More especially, as more of the secrets of the sky were revealed and categorized, and on Earth the power of the waterwheel brought nature increasingly under control, there grew the feeling that man was the measure of all things.

The theory of the four humours goes back to Hippocrates, the fifth-century B.C. Greek physician. According to this theory the temperament of a person depended on which of the humours was dominant in his character.
Top left: The man with wet, black bile was pessimistic and loved sad songs.
Top right: Dry, yellow bile made a man choleric.
Bottom left: Hot, dry blood made for a sanguine, optimistic character.
Bottom right: The lazy, phlegmatic man owed his character to cold, wet phlegm.

127

The convergence of experience in the motive power of water with an increasingly scientific view of the rotating heavens was to lead to the development of one of the most fundamental inventions in the history of technology: the mechanical clock. Interest in knowing the exact hour had begun with the foundation of the Benedictine Order in the fifth century, when, in his Rule, St Benedict had specified certain activities at certain times of day and night: at midnight, 3 a.m., dawn, sunrise, *tertia* (halfway between sunrise and midday), *sexta* (midday), *nona* (mid-afternoon), vespers (an hour before sunset) and 9 p.m. These hours, associated with prayer and ritual, he called 'canonical' hours. The saying of prayers at the right time was part of the discipline Benedict imposed on his Order, and was essential for the monks' salvation. During the day, the canonical hours could be determined easily, with a sundial, but at night they required a member of the community to sit up all night and ring the bell to wake the others at the right time. The drive to find an automated timekeeper must have been strong. The first chronicled evidence we have that the Church was using some semi-automatic device to tell the time comes from a monk called Jocelyn de Brakelond, who wrote, among other things, about the fire in the great church at Bury St Edmunds in Suffolk on 23 June 1198. The church was a centre of pilgrimage as it was supposed to house the body of the saint, and a fire was a potential disaster – both spiritually and financially – for the community. On the night in question the monks were awakened by the crackle of flames, and, in Jocelyn's words, 'We all of us ran together and found the flames raging beyond belief and embracing the whole refectory and reaching up nearly to the beams of the church. So the young men among us ran to get water, some to the well and others *to the clock . . .*' (my italics). The reason they put fires out with clocks is that the clocks were water-powered.

By the twelfth century, then, the church was using the water clock, or *clepsydra* (from the Greek, meaning 'water stealer'). This type of clock must have been introduced into northern Europe because of the number of cloudy days on which the sundial would have been useless. The clepsydra probably originated in Egypt, where it may have been used to regulate water supplies, since, by the time the Arabs came to use it, they were still measuring water with a remarkably similar device to the clock. The idea was to fill a bucket with water and pierce a hole in the bottom. The farmer receiving his water supply would be given water until the bucket ran out, whereupon he would pay for it in 'bucketsful' of time. At some unknown point during the twelfth century the problem of waking up at night at the right time to ring the bell was solved, because, in a new version of the Rule for Cistercian monks, it says that the sacrist 'roused by the sound of the clock

The water clock operated by means of a reservoir (top left) with a constant supply of water, kept at a steady pressure by means of an overflow outlet. This reservoir dripped into a main tank (centre), which had a baffle wall to stop ripples and a float carrying a pointer. As the water level rose, the pointer indicated the time passed on a scale representing the hours. This was graded to allow for the difference in length of hours in winter and summer – those shown on the left of the cylinder apply to the longer hours of summer.

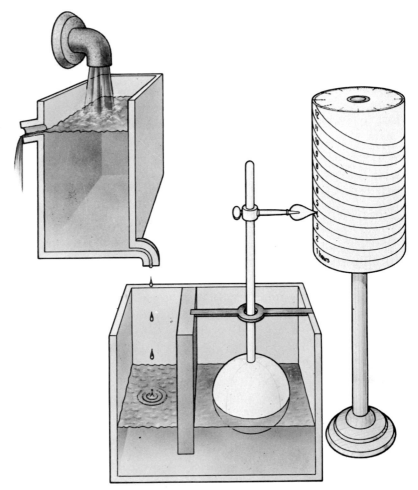

shall ring the church bell'. And on a slate found near the Abbey of Villers, near Brussels, and dated around 1268 is another timekeeper's rule: 'You shall pour water from the little pot that is there, into the reservoir until it reaches the prescribed level, and you must do the same when you set the clock after Compline [the evening service] so that you may sleep soundly.' The earliest reference to a mechanism operating an alarm clock comes from an eleventh-century manuscript found in the Benedictine monastery of Santa Maria de Ripoll, at the foot of the Spanish Pyrenees, which describes such a clock.

From all the evidence it seems clear that the monastic clock-makers were not interested in telling the time as we would be today, but merely in sounding a bell at the right moment in the passage of the day or night. The principal reason for this difference of approach to time is that the length of their hours changed according to the season. In winter the day is short, in summer, long. In consequence, the length of the intervals between the canonical hours lengthened and shortened in summer and winter. So the development of a mechanical clock telling regular hours was for the monks a problem – one they could only solve by introducing devices for setting the clock to ring at longer or shorter intervals according to the time of year. These were pegs, inserted in holes on the clockface, which tripped the alarm. By 1271 Robert the Englishman, teaching at Paris or Montpellier, wrote: '. . . clock-makers are trying to make a wheel which will make one complete revolution for every one of the equinoctal circle [the day], but they cannot quite perfect their work . . . the method for making such a clock would be to make a disc of uniform weight . . . then a lead weight should be hung from the axis of that wheel and this weight would move that wheel so that it would complete one revolution from sunrise to sunset . . .' The interest in trying to make a mechanical clock sprang from two sources: the need to tell the time with a system which would not freeze in winter, as the clepsydra did, and the desire (born of the new astronomical knowledge and the magic of astrology) to recreate what happened in the sky on a controllable, human scale, and bring God's clockwork down to Earth. Robert had put his finger on the problem. It was easy enough to spin shafts and dials with lead weights and cords. The question was how to regulate the speed of the spin? Nobody knows where the answer, the 'verge and foliot' system came from, but it is one of the most ingenious of man's inventions. It worked by holding back the speed at which the main shaft turned as the weight wound round it fell.

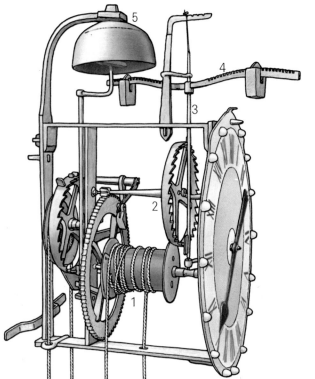

The first clockwork. The main weight-driven wheel (1) was held in measured check by the rotation of a toothed wheel (2) controlled by a vertical verge (3), with blades fixed to it so as to catch in the teeth of the wheel at top and bottom alternately. The pressure needed to hold back the turning wheel was provided by the weighted, T-shaped foliot (4), whose momentum gave the verge blades just enough inertia to resist the wheel for a moment. As one blade was kicked away by the teeth, the other engaged. The mechanism could be used to trigger an alarm (5).

The famous Horloge de sapience *illustration (c. 1450). On the left is a mechanical clock with a single hand and a 24-hour dial. Other timekeepers are on the table, right: a spring-driven clock, a quadrant, a sundial.*

The verge and foliot system was to have the most profound effect on the society into which it came. What had begun as a machine for telling monks when to pray, rapidly became a regulator of every aspect of human life. The new clock was probably in action at some time round 1280 or 1290, and was used at the beginning merely for striking the hours. The early clocks placed in church towers had no dials or hands, and at this stage probably still served to tell the time of day to the priests rather than to the community at large. Initially the new mechanism was used by ecclesiastical astronomers in the form of automated calendars. The earliest of these appears to have taken the form of a large face on which a pointer showed the signs of the zodiac, while windows showed other parts of the mechanism rotating the phases of the moon, the position of the sun, the major constellations as they rose and set, and certain dates, principally those of the feast days. This last was the most important function of the new clock, since the Church had a considerable number of feast days whose date depended on astronomical data. The only movable feast left today is Easter, whose date depends on the phases of the moon. The feast day was a matter of concern both to the priest, whose salvation depended on its observance, and to the populace at large, whose life revolved round these days. Sowing, harvesting, market days, holidays and major religious rituals were all held on particular feast days. It was this connection with work that was to take the clocks out of the hands of the Church and into the town squares all over Europe.

There are various claims as to where the first clock was set up, since none of the documents give mechanical details, so there is no way of knowing whether or not a 'clock' referred to is mechanical. There was some form of clock in Westminster in 1288, and one in Canterbury cathedral in 1292. One is mentioned in Paris, in 1300. In 1321 Dante referred to a clock in the *Paradiso* in such a way as to suggest that his readers would have been familiar with its mechanism. The first clear reference to a mechanical clock is to one in the Visconti Palace in Milan, in 1335: 'A heavy hammer strikes 24 hours . . . distinguished hour from hour, which is of great necessity to all conditions of men.' The writer was echoing the thoughts of those in civil power, because, as the clocks spread during the fourteenth century – to Wells, Salisbury, Strasbourg, Paris, Bologna, Padua, Pavia, Ferrara – so did the realization of their economic value. In 1370, in Paris, Charles V ordered all the bells in the city to be rung at the sound of his new clock in the tower of the Royal Palace, so that all men should know the king's hour. Clocks indicated the time for opening and closing the city gates, for the beginning and ending of curfew (the putting out of fires, a matter of public safety in towns built predominantly in wood), for the hours of the watchmen at night, and above all for the hours of starting and stopping work. The citizens, or at least those of them with a stake in the town's prosperity, responded eagerly to the opportunities a clock offered. The town council of Lyons received a petition for 'a great clock whose strokes could be heard by all citizens in all parts of the town. If such a clock were to be made, more merchants would come to the fairs, the citizens would be very consoled, cheerful and happy, and would lead a more orderly life, and the town would gain a decoration.'

As clock-making techniques improved it became possible to build clocks small enough to be used inside the private rooms of their owners,

A German portable alarm clock (c. 1550) in gilt brass, with a steel and brass movement. The dial gives both the usual twelve hours and the Italian twenty-four hours. Watchmakers in Nuremberg were only admitted to the guild after making such a clock, as well as one which carried an automated calendar and showed the phases of the moon.

and these new chamber clocks rapidly became unparalleled status symbols. Such was the great clock made by Giovanni de' Dondi, a professor of medicine and astrology, for his patron the Visconti Duke of Pavia – where it stood in the castle library, the wonder of Europe. It was finished in 1364, and consisted of 297 parts held together by 305 pins or wedges. (The idea of using the screw for holding things together, rather than merely for pressing things, was yet to come.) Dondi's clock recorded the hours, the minutes, the rising and setting of the sun, the length of the day, the Church feasts, the days of the month, the trajectories of five planets, the phases of the moon, an Easter calendar, star time, and an annual calendar. It did all this with extreme accuracy. The weighted balance – his equivalent of the foliot – swung back and forward 43,000 times a day, giving the clock a two-second beat. Unfortunately Dondi's masterpiece was later destroyed in a fire, though detailed blueprints were left, copies of which may be seen today in the Smithsonian Institution, Washington, and in the London Science Museum. The problem with all these clocks was what time they should tell. Some places, England among them, decided early on that a mechanism for striking twenty-four separate strokes was too complex and costly, and settled for the repetition of a twelve-hour strike. Most of the rest of Europe gradually accepted this point of view, save Italy, which retained the twenty-four-hour system until the nineteenth century. One of the most complicated systems was adopted for a time in Nuremberg, where clocks were built to tell the time from sunrise to sunset, with inbuilt mechanisms to handle the varying length of the days throughout the year.

The value of the clock was not lost on the merchant and businessman. By the middle of the fifteenth century, as Europe was recovering from the effects of the Black Death and the economy was beginning to boom once again, there was considerable demand for a clock that was portable. Time was becoming money, and around 1450 an offshoot of water power took the development of the clock one step further. At the time, the new water-powered bellows were operating in blast furnaces, and waterwheels were powering mills to crush the ore to feed them. The amount of metal in circulation increased dramatically, and the more it increased, the greater was the demand for it. Metal-working skills spread, concentrating in those areas closest to the mines from which the ore was extracted. One of these centres was Nuremberg, in the mountains of southern Germany, and it was probably there that a craftsman in metal – perhaps a locksmith, or an armourer – realized that sprung metal could be used to replace the weight drive. One of the earliest examples of the new spring-driven clock is shown in a portrait dated around 1450 of a Burgundian nobleman, with his new clock proudly displayed in the background.

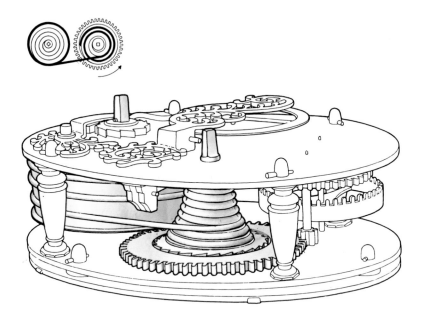

Its works reveal the presence of the mechanism developed to solve the problem which the spring created almost as soon as it was first used: that, as a spring unwinds, its power weakens. As a driving force the spring was useless, since as it began to unwind the clock would go fast, and as it weakened the clock would go slower and slower. The device invented to counter this difficulty was known as the fusee, taken from the Latin word for a spindle.

By the middle of the sixteenth century the spring-driven clock had decreased in size to the point where it had become a watch. These small clocks and watches were adequate for the needs of the vast majority of their users. They were not, however, adequate for the astronomers. As their astrolabes, quadrants, planetaria and other tools of observation became bigger and more precise the astronomers began to complain that the spring-driven clock was not accurate enough. The quality of the metal varied, and the behaviour of the spring depended on factors such as temperature, age and lubrication. Some of these clocks were out by as much as four minutes a day, and for close examination of the regular movement of the planets, four minutes was too much. This need for greater accuracy was brought to a head in a roundabout way by a Dutch spectacle-maker, Hans Lippershey. On 2 October 1608 he offered an invention to the Dutch Government for use on the battlefield. It was called a 'looker' – we should call it a telescope. The Dutch Army reformer, Prince Maurice of

The spring-driven clock, using a fusee as a control mechanism. The upper diagram shows the spring wound inside a barrel to which it is attached. As the spring unwinds, it turns the barrel, pulling on a gut cord fastened to the outside of the barrel. The problem of the strength of the spring lessening as it unwinds is solved by causing the gut to turn a cone-shaped fusee (the main clock drive). The spring does the hardest work, turning the narrowest end of the fusee, when it is still tightly wound. Its work becomes easier as it weakens, and the gut unwinds towards the broad end of the cone.

Nassau, was keen to try any device that would help modernize his military tactics, and he took the new telescope out for field trials; as a result Lippershey was awarded a prize of 900 florins, on condition that he make his invention binocular.

It may seem odd that a period of over three hundred years separates the first use of glass lenses in spectacles – attributed variously to an unknown inventor in Pisa and another in Venice, around 1300 – and the use of such lenses in the telescope. This is another example of the connection between invention and social need. As the European economy picked up after the centuries of the invasions (the 'Dark' Ages) any device that would prolong the working life of ageing scribes was to be welcomed. But there was no demand for the telescope during this period, which was prior to the invention of gunpowder and the use of cannon on the battlefield, when the view of the universe precluded the existence of planetary bodies as three-dimensional, observable phenomena. This is why the moment of invention is so often identified with the moment in which the artefact comes into use. In many cases there are times when an invention is technologically possible – and in which it may indeed appear necessary, as the telescope may have – but without a market the idea will not sell, and in the absence of the technical and social infrastructure to support it, the invention will not survive.

The telescope is a particularly good example of this principle. When it appeared, astronomy had reached the point of questioning the old Ptolemaic view of the universe as a series of concentric spheres, each carrying a planet, or a constellation, at the centre of which stood the Earth. Observation through the use of astrolabes had shown that there were anomalies in the supposedly perfect regularity of the heavenly movements. The planets, for instance, did not behave like objects circling at a uniform distance from Earth, and by the sixteenth century theories had already been developed to show that their orbits were not perfectly circular. These theories suggested that the only way to explain the anomalies lay in an entirely different concept of the universe: to see it as a system with the sun at the centre, and the planets, including the Earth, circling it. This would account for the moon as a satellite of Earth behaving as the Earth itself did, or Venus, as a satellite of the sun. An example showing another of the planets as a satellite-bearer would add immeasurable strength to this theory, and in 1609, when Galileo first looked at the sky through one of the new telecopes, he saw that evidence. There were satellites circling Jupiter. Moreover, the moon appeared to be like the Earth, not a perfect sphere as the old, established view had held, but a world with mountains and valleys on it. And there were infinitely more stars in the sky than had ever been dreamed of.

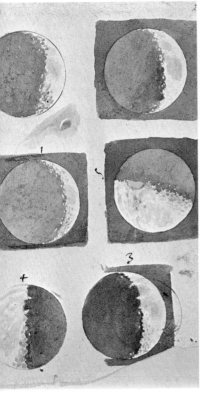

Galileo's wash drawings of the phases of the moon observed through his telescope in 1610. The irregularity of the lunar surface destroyed the theory that all planets were perfect spheres.

The telescope radically altered man's view of his position in the universe. No longer was he the centre of things, the uniquely chosen manifestation of God. The telescope freed man to look at himself and the world about him. Since nothing was certain any more, and the 'new philosophy called all in doubt', everything could and should be investigated.

The Jovian satellites created further problems for the clock-makers. Jupiter's satellites swung in orbit round the planet, and the time of these orbits had to be measured accurately if the new discoveries were to be inserted into the body of knowledge in any significant way. As has been said, the spring-driven clocks were hopelessly inadequate. Myth has it that the problem was solved by Galileo himself, observing a lamp swinging in Pisa cathedral and realizing that the timed length

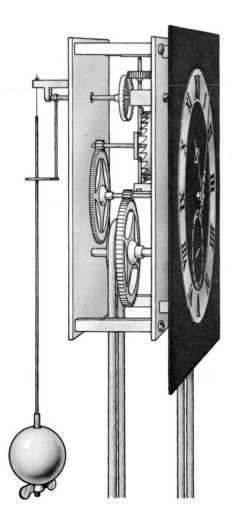

The Huygens pendulum clock. As can be seen, the system uses the regular swing of the pendulum in the same way as the foliot had done, to give inertial power to the verge blades holding back the toothed wheel which regulated the speed with which the main clock weight fell.

of the swing was the same whether the lamp swung in a wide or a narrow arc. So the pendulum effect could be used to provide a time measurement of exact length. From his writings it appears that Galileo went no further than to design a pendulum device which would operate a mechanism to move a toothed wheel on, one tooth at a time. It was a Dutchman, Christian Huygens, who made the first pendulum timekeeper.

The Dutch were particularly interested in timekeeping because they were in the early phases of opening up the East, and their navigators had come across a problem which is only found when a ship sails east or west for any appreciable distance. Huygens' work was partly in response to this need. The pendulum clock provided the answer to the astronomers' demand for exact timing, because it used the unvarying length of the pendulum swing in much the same way as the weighted foliot had been used – to act as a check on the movement of the toothed main wheel of the clock. The pendulum did so with very much greater accuracy, but this did not solve the navigators' problem, since a pendulum clock cannot be used in a pitching sea. The reason sailors bound east and west needed to know the time was that the star tables they carried told the navigator what position the moon or sun or a star should be in the sky at a certain position on Earth at a certain time of day or night. These tables had always been computed on a meridian: the north–south line that ran through the place where they had been computed. Thus, a ship sailing north or south on the Greenwich meridian could use the tables to find out how far north or south of Greenwich the ship was. Once the ship moved east or west, however, the positions of the stars as catalogued by the tables would differ by the longitudinal (east or west) distance of the ship from its home meridian. Since the Earth turns once every twenty-four hours, it follows that for every 15 degrees east or west, sunrise will be one hour earlier or later. The position of the stars will be similarly altered. Unless the navigator had an accurate clock to tell him what time it was at his home meridian, he could not make the necessary adjustment to work out his position. Each minute that the navigator's clock was fast or slow would give him a position that was, at the Equator, inaccurate by about 15 miles. In a world increasingly dependent on overseas trade, for a ship to miss landfall by 15 miles was economically unacceptable. It became imperative to find a clock that would tell the time accurately over long periods of time, in conditions such as those encountered on transoceanic voyages. There appeared to be only one answer: to improve the quality of the metal from which springs were made. The man who made that possible, and also helped to bring about the Industrial Revolution, was a clock-maker from Doncaster, in northern England, called Benjamin Huntsman.

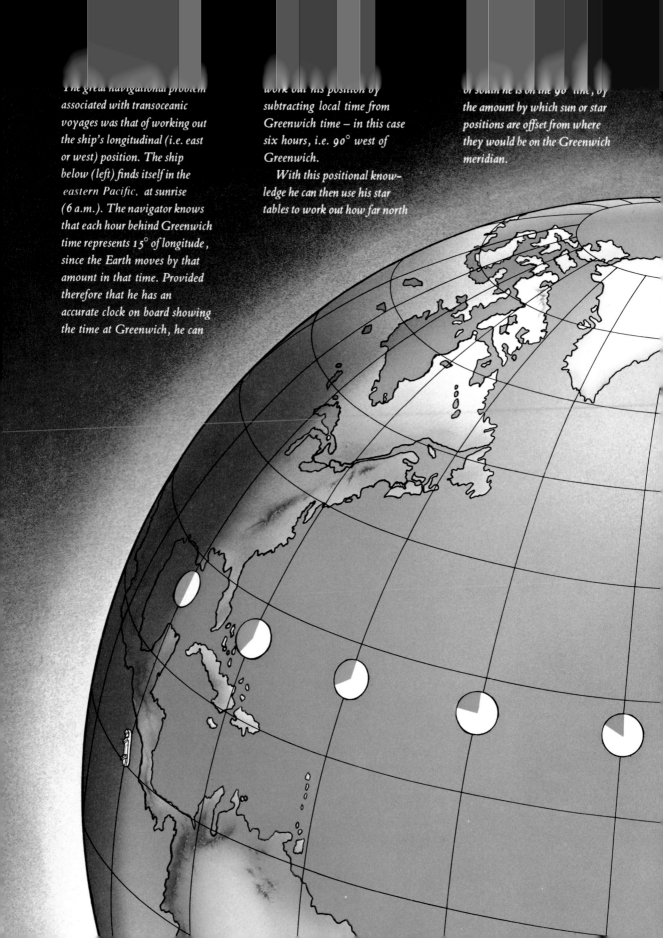

The great navigational problem
associated with transoceanic
voyages was that of working out
the ship's longitudinal (i.e. east
or west) position. The ship
below (left) finds itself in the
eastern Pacific, at sunrise
(6 a.m.). The navigator knows
that each hour behind Greenwich
time represents 15° of longitude,
since the Earth moves by that
amount in that time. Provided
therefore that he has an
accurate clock on board showing
the time at Greenwich, he can

work out his position by
subtracting local time from
Greenwich time – in this case
six hours, i.e. 90° west of
Greenwich.

  With this positional knowledge he can then use his star
tables to work out how far north

or south he is on the 90° line, by
the amount by which sun or star
positions are offset from where
they would be on the Greenwich
meridian.

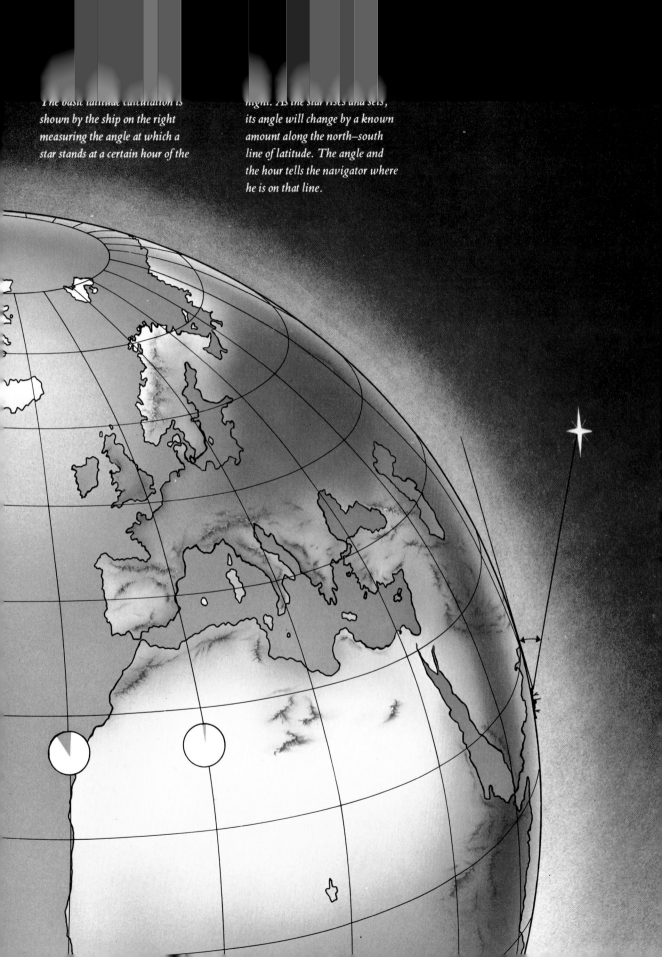

The basic latitude calculation is shown by the ship on the right measuring the angle at which a star stands at a certain hour of the night. As the star rises and sets, its angle will change by a known amount along the north–south line of latitude. The angle and the hour tells the navigator where he is on that line.

The process by which fundamental change comes about at times has nothing to do with diligence, or careful observation, or economic stimulus, or genius, but happens entirely by accident. There were hundreds of clock-makers like Huntsman all over Europe who were equally dissatisfied with the quality of the springs in the clocks they were making. Many of them must have cast about for the answer to their dilemma, but nothing suggested itself. Everywhere, the technique for making steel at the time was the same: alternate layers of charcoal and iron were piled up, covered with a layer of fine sand, and kept red hot for several days. During this time the carbon in the charcoal diffused into the iron, forming a surface layer of steel which was then hammered off. Many of these layers were then hammered together to produce layered, laminate steel: good enough for knives, but liable to snap or deform when bent into springs. Huntsman happened to live near a glass-making community, and at a time when Abraham Darby had discovered the high temperatures that could be obtained with coke. The glass-makers were using coke to fire their ovens, and lining the ovens with Stourbridge clay from local deposits. This clay reflected heat back into the ovens, raising their temperature even further. Huntsman also saw that the furnace men mixed their raw materials for making glass with chips of old glass, which because of the high furnace temperatures would become molten and run together with the freshly made glass.

Huntsman's steel furnaces, just outside Sheffield, in Yorkshire. These works gave birth to the Sheffield steel business only after Huntsman had convinced the cutlers of rival French interest in his process.

In 1740 Huntsman set up in the village of Attercliffe, making steel by the same method. Using chips of laminated steel in his furnace crucibles, he was able to melt the different layers into a homogeneous fluid which when cooled became strong and highly tensile. This new 'crucible' steel was eventually to form the springs of the ship-borne chronometers of the late eighteenth century, accurate to within seconds over periods of several months. But Huntsman's steel was to have an even more profound effect on the eighteenth-century world. At about the same time that he started producing steel, the son of a French immigrant to England, John Dolland, had been experimenting with two lenses in a telescope. He was attempting to prove Newton's theory that the use of more than one lens would break up the light into its prismatic colours, and in doing so he found that Newton was wrong. What happened to the light, he discovered, depended on the type of glass used, so he tried using crown glass (the type used in windows) for one lens, and flint glass (the type Huntsman had seen being made) for the other. Light came through both lenses un-impaired, and Dolland patented his achromatic lens in 1757. It was taken up immediately for use at sea, on a variant of the quadrant which John Hadley had developed in 1731 for more exact star sighting.

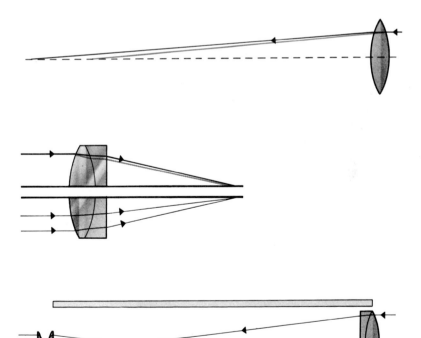

The evolution of the telescope.

Newton discovered that light passing through the two surfaces of a lens would split into two colours.

Dolland's solution to the problem: by passing light through two different kinds of glass (and therefore through four surfaces) the separated colours would be focused on the same point, as would the image of the object (lower half of lens).

In a telescope, this results in both image and light passing through the lenses to emerge integrated.

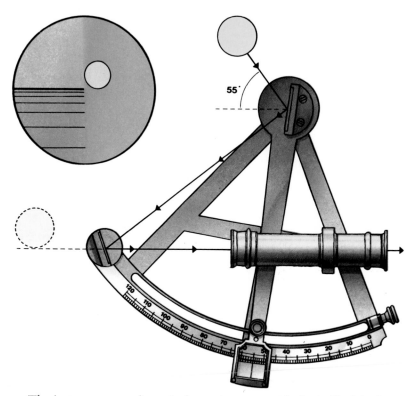

55°

120 110 100 90 80 70 60 50 40 30 20 10 0

The instrument used a twin-lens telescope, with the half of the lens further from the eye silvered over. When a star was sighted through the telescope, a pivoted mirror was swung until it reflected the star in the silvered half of the lens while the observer saw the horizon through the clear half. The angle to which the pivoted mirror had to be swung for this to be possible was the angle at which the star stood in the sky. The accuracy of this sextant was such that a navigator could position his ship to within a hundred yards. The trouble with the new telescopic sight was that it was too accurate for the scales available at the time, marked out on the metal as they were by hand. Some new way to mark highly accurate divisions of degrees down to one-sixth degree (i.e. 10 minutes of arc) had to be found.

Until then the divisions had been marked with dividers, by hand. On large-scale instruments this did well enough, but with the telescopic sight it was out of the question. To produce marks at every 10 minutes of arc round the circumference of a small instrument meant scribing a total of 2160 marks to complete the 360 degrees. The pressure to solve this problem was strong, and further stimulated by government prize money – including the big prize of £20,000 for the man who could find a way to measure longitude accurately (to one-third degree). In 1739 a Yorkshire clock-maker called Henry Hindley, who had gone as so many clock-makers had into the business of astronomical instruments, adapted a clock-wheel cutter for marking

Hadley's sextant. The circular inset (top left) shows what the observer sees when the sighting is made and the image of both sun and horizon strike the eye. The graduated scale along the bottom of the instrument gives the angle of altitude indicated by the pointer, which is set in the frame at the foot of the central pivoting bar on which the mirror is mounted.

The dividing engine (below, right) consisted of a bronze toothed wheel, cast in one piece and reinforced by spokes. The plate to be divided was laid on top of this wheel, which was set horizontally on three rollers so that it revolved around a vertical steel axle with great precision. The gearing principle of the circular dividing engine was provided by the tangent screw (below), which advanced the large wheel by an exact amount at each turn.

divisions. The principle was simple: a circular plate with teeth in it lying horizontally acted as a base plate. At a tangent to its edge was a small screw. As this screw was turned its spirals caught in the base plate teeth and turned it very slowly. By counting how many turns of the screw it took to turn the plate 360 degrees, the operator knew how many it took to turn a degree, or a minute. The chances are that Henry Hindley was over-generous with his ideas, because in all probability it was one of his workers, John Stannicliffe, who took the fame due to his master by passing the principle on to a man already well established in influential circles. This man was Jesse Ramsden, and he was married to John Dolland's sister Sarah. In 1774 the Ramsden dividing engine brought Jesse £300 prize money, and a lot of work. His prizewinning engine could finish a whole sextant in less than twenty minutes, and cut the cost of dividing instruments by as much as ten times. It had the required 2160 teeth, and each one of these was turned by only one turn of the tangent screw. To give Ramsden his due, he knew where to go in order to get what he needed: Yorkshire for the engine – France for the screw. Ramsden's problem was that the screw which turned the base plate had to be extraordinarily accurate. Fortunately for him, a French instrument-maker called Antoine Thiout had developed a way of making lathes more accurate, and it was this French principle that Ramsden latched on to.

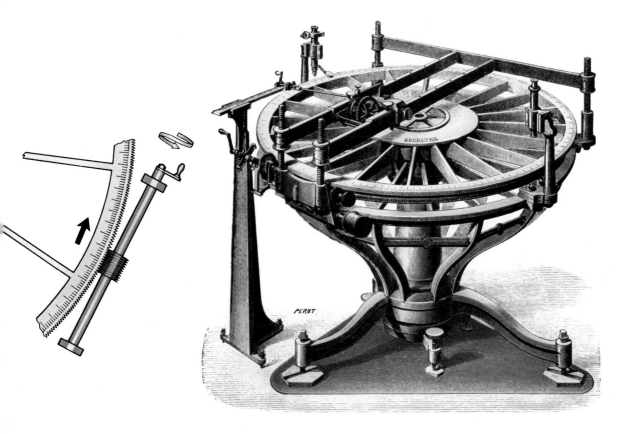

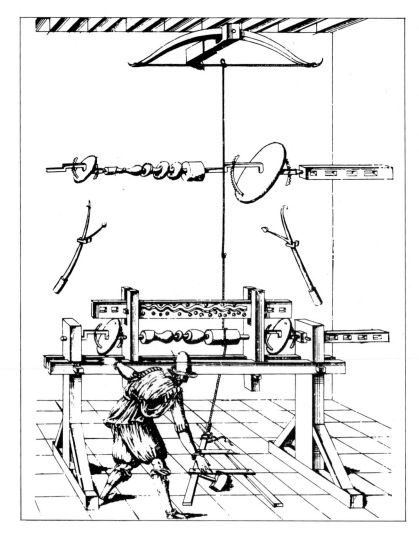

A variant on the pole lathe, operating on the power of a bow. When the foot-pedal is released, the bow pulls the cord up, spinning the work piece. Note the template above the work piece which the turner uses to guide his knife so as to produce the desired shape.

    The lathe had been in use since ancient times, but by the eighteenth century it had come a long way from the pole drive. This had used the spring in a wooden branch or pole to drive the lathe. A rope from the pole was wound round the piece of wood to be worked, and was then attached to a pedal. As the pedal was pushed down, the pull on the rope caused the work piece, mounted horizontally, to spin one way until the pedal was fully depressed, at which point the spring in the pole would pull the rope upwards, and the work piece would spin in the opposite direction. To cut the piece, the worker would steady his knife against a wooden block and cut on every other spin, i.e. when the wood was turning in the same direction. With the arrival of water power the drive became continuous, in one direction. By the seventeenth century lathes were equipped with fancy gearing systems and cams to permit complicated patterns to be shaped (which may account for the particularly bulbous furniture of the period). But with the increasing use of iron, traditional lathes became inadequate, because since iron was so hard the operator needed more support for his knife than his hands. The turners of ornamental furniture had developed a tool-rest which brought the cutting knife in from the side, mechanically, against the piece to be worked. The problem of controlling the work longitudinally – that is to say getting an accurate and even cut all the way along the piece as the knife moved along its length – was solved by Thiout, working on cutting fusees for clocks.

The idea was to use a screw as a moving base for the knife: if the knife were mounted on a support with a threaded hole through it and a screw fitted to that hole, when the screw was turned the support carrying the knife would travel along the screw, and could thus be moved accurately along a work piece. This 'lead screw' idea also solved Jesse Ramsden's problem with his tangent screw. He cut a lead screw in exactly the shape he wanted his tangent screw to be, so that when the lathe operator turned the lead screw the knife cut the same pattern in the work piece, and there was Ramsden's tanget screw, every time. And it is at this point that Huntsman comes back into the story, because only one metal would cut every kind of iron with ease – Huntsman's crucible steel. This put the edge on the Industrial Revolution, because without it the lead-screw lathe would have been a good deal less effective, and without the ability to cut metal very precisely the Industrial Revolution could not have taken place.

The man who gave the machine-makers a perfect synthesis of all that was known about cutting metal, in one machine, was called Henry Maudslay, and had it not been for a series of widely publicized robberies in London, he might have stayed where he was at the time, in the smithy at the Woolwich Royal Arsenal. The robberies coincided by chance with the marketing of a new and very much superior lock by a lavatory-maker called Joseph Bramah (who, himself, would have been a farm boy but for an accident in youth which made him turn his hand to carpentry and start making lavatories, from which he progressed to locks). In 1789, when Maudslay was eighteen, Bramah hired him as an apprentice locksmith. Over the next eight years, as Bramah himself confessed, it was Maudslay who made the mass production of Bramah's lock possible, by devising tools that would help the locksmiths do their work faster and with more accuracy. But in 1797, at the age of twenty-five and with a wife and family to support (he had married Bramah's housekeeper), Maudslay asked for a rise on the thirty shillings a week he was getting at the time. Bramah refused, and Maudslay walked out, to start his own workshop just off Oxford Street, in London. He took with him the designs for the lathe he had already developed.

The famous Bramah lock remained unpicked for 40 years, when an American locksmith, Hobbs, succeeded after 16 days' continuous effort.

The difference between Maudslay's lathe and all the others that preceded it – and there were many – was that he had scaled up the old lathe used for ornamental work to industrial size, without losing any of its precision. He managed this because he was a near-fanatic for accuracy. The 1800 Maudslay lathe (known at the time as the 'go-cart') worked on the lead screw principle, with the sliding tool-rest perfectly mounted on accurately planed triangular bars. Maudslay's reputation rested on the fact that with his machines he could produce more accurate work faster than anyone else. He also spent time chatting to a Frenchman who passed his shop in Wells Street every morning and this chance encounter brought Maudslay into contact with a fellow Royalist émigré from Republican France, Marc Isambard Brunel, who had recently returned from America.

Brunel came to Maudslay in 1800 with an idea for making ships' blocks. At the time, these blocks were essential to the running of a ship. They were (and still are) shaped wooden blocks with a pulley, or two, set into them, so that a fixed line attached to the block could be tightened or loosened at will. They were used for control of the rigging, for hauling any heavy material, and for moving guns into and out of position rapidly. It took over 1400 blocks to operate a 74-gun ship – a vessel, incidentally, only of the third class. In 1800 these blocks were made by hand. Brunel had an idea for making them by machine, and he wanted Maudslay to make the machines.

It took Maudslay six years, and the first blocks were being made in 1808. It was the first, large-scale mass production unit in the world, and although the Portsmouth blockmaking yard did very well, the idea of extending the system to other industries did not catch on in England for another fifty years. By that time the industrial lead it could have given England was lost to the country where Brunel claimed to have had the idea in the first place: America.

Right: The Brunel–Maudslay blockmaking machinery. With it, ten unskilled men could make 130,000 blocks a year. The illustration shows the principal stages of manufacture, with (below, left) the pulley before insertion into the block, and (right) the pulley in operation, reducing effort by a factor of three.

Below: Maudslay's lathe. The diagram shows how the turning lead screw (below) carries the knife gradually along the work piece, cutting a screw whose thread could be varied by the gearing at the right-hand end of the lathe. Maudslay also used a fine screw as the basis for a micrometer with which he measured accuracy to 1/1000 of an inch. The principle on which it worked was the same as Ramsden's tangent screw.

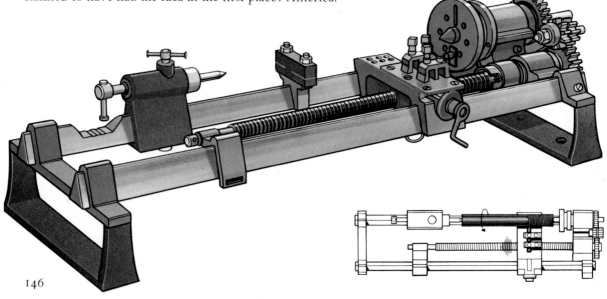

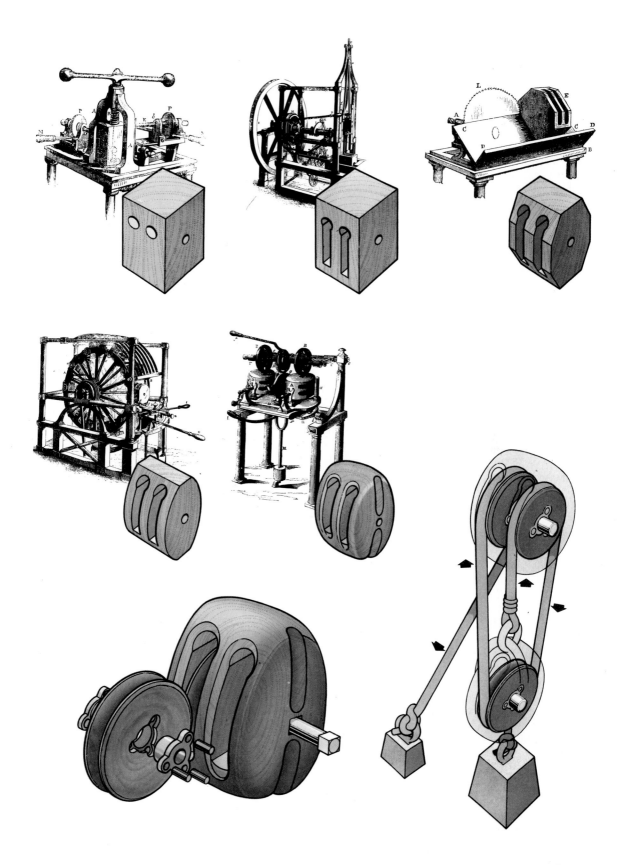

Automated machines that would take work away from skilled men failed in England for very understandable reasons. By the early nineteenth century all the water power, in the form of rivers, and all the raw materials, such as coal and wood, had been in private hands for centuries. For every skilled experimenter like Maudslay there were a thousand whose only security lay in their craft, acquired through years of apprenticeship and workshop experience. In a country as geographically and socially limited as England, there was nowhere else for these men to turn for income, save to lose all their skill had gained them and become common labourers. Automated machines represented a threat to their livelihood, and they opposed them. However, for some years many of these skilled men had been defying the law to seek greater opportunity in America, a country where none of the English limitations on a man's expectations existed. As conflict developed with the colonists, the British Parliament had enacted legislation to prevent any transfer of technology across the Atlantic. The law of 1750 had read: 'No mill or other engine for slitting or rolling of iron, or any furnace for making steel shall be erected . . . in His Majesty's Colonies of America.' The penalty for infringement was £200, more than a skilled man could save in a lifetime. After the War of Independence, the law was amended to read: 'no export of tool, engine or persons connected with the iron industry . . .' The laws did not prevent considerable numbers of men from emigrating, dressed as common labourers, or even as women, with their tools camouflaged in boxes – even, on occasion, marked 'fruit trees'. The greatest attraction towards America was for the men who made machines for the textile industry. From the earliest days of the Republic it had been the practice for local citizens in places like Philadelphia or Providence, Rhode Island, to go to the quays on the arrival of ships from England, seeking men with these skills and offering them employment and bonuses to take up their trade in their new home. With the embargo on trade of 1807–9 and the war with England in 1812 the textile trade boomed, as the political situation diverted vast amounts of money from overseas investment into the domestic market.

The textile industry in America rapidly took to the idea of mass production for several reasons. First the building of factories, and their siting close to sources of water power, was easily done in America. New England was covered with forests, and entire rivers could be bought for their water rights. Then there was the problem of labour. With the government encouraging settlement by selling land to immigrants at $1.25 an acre, most men headed out into open country and became farmers on sections of land they could never have afforded at home. The only readily available pool of labour was the farmers' daughters. In some cases these girls simply wanted to get away from the drudgery of the farmhouse; more often they had a brother's education to finance, or a dowry to collect. The unskilled nature of these girls forced the entrepreneurs to opt for automated machinery that needed minding, rather than skilful operating. In 1813 a Boston financier called Francis Cabot Lowell set up the Boston Manufacturing Company at Waltham, near Boston. He had previously spent time in England, where in Manchester he had observed cotton manufacture at first hand. When he returned to America, just before the war of 1812, he met an expert machinist called Paul Moody who was to build most of his machines. In 1814 he hired a recent arrival from England who had brought with him the secret of the new power loom, and the way was clear for a textile version of the Portsmouth block-making system. Lowell hired farm girls to mind the machines, and then put the whole process under one roof, powered from a central source. Raw cotton went in at one end of the factory and woven cloth came out at the other. The first stage in what was to become known as the American System of Manufacture was complete. The second stage was not long in coming, and it too was born of restrictive practices on the part of European workmen.

The Lowell Mill girls at work. During their brief leisure hours they lived in purpose-built dormitories and followed a strictly enforced routine, designed to ensure their moral welfare.

In the middle of the eighteenth century a French gunsmith called Honoré le Blanc had worked out a system for making gun parts to a standardized pattern, so that if a part broke it could be replaced by another part that would fit the gun exactly. Since up to that time the manufacture of muskets had been in the hands of craftsmen, these latter were unwilling to adopt new methods that would reduce their status or employment. So Le Blanc turned to Thomas Jefferson, the American Ambassador to France, with the idea. Jefferson saw at once that such a system would rid America of her dependence for arms on those states with which she had been and might again be at war. He tried and failed to persuade Le Blanc to go to America. But he wrote enthusiastically to the Secretary for War about the idea, and when he returned to America he set about persuading Congress to adopt the principle in the nation's armouries. He had an ally in George Washington, and by 1798 a contract had been given to Eli Whitney, the self-styled inventor, for 12,000 muskets made by the new system –

Eli Whitney's arms factory at New Haven, Connecticut, in 1820. The houses in the background are for the workers and their families.

a method which, by the way, Whitney claimed for his own. Whether Whitney was the first even in America to copy Le Blanc's idea is a matter of doubt. There were two other gunmakers, John Hall and Simeon North, producing guns by the same method at the same time. Be that as it may, Eli Whitney continues to be known as the inventor of the system, because there is a need for heroes. The machines operating in the armouries produced parts that were standard enough to be interchangeable, and by 1815 Congressional contracts stipulated this quality in the muskets, pistols and rifles they paid for. During the next fifty years the factory system of Lowell and the interchangeable parts of the armourers combined in the American System of Manufacture, with factories beginning to turn out quantities of goods made of interchangeable parts, such as sewing machines, locomotives, bicycles, and the product of much of the bicycle manufacturers' experience – the motor car. Cadillac was first a bicycle, then a car.

As America was opened up, and the vast natural resources of the continent revealed themselves, the American System of Manufacture had put the country ahead of the rest of the world in terms of manufactured output by the end of the nineteenth century. The new machines produced goods for everyone. The machines became specialized, rather than their operators, and the American eagerness to use them was reflected in a need no other country had: to absorb thousands of non-English-speaking immigrants who in most cases came from pre-industrial societies. It was partly in answer to the problems created by these workers that work began in 1881 at the Midvale Steel Plant in Pennsylvania, under Frederick Taylor, an engineer, to attempt to systematize the way a job was done. Taylor used a stop-watch to time and analyse the separate movements involved in production. At the same time another engineer, Frank Gilbreth, and his psychologist wife Lilian started using the new motion pictures to take slow-motion films of bodily movement in order to reduce them to a standard, efficient minimum. The time and motion studies of Taylor and the Gilbreths finally made the worker a part of the American System, as interchangeable as the machines he operated. The genesis of the modern production line was complete.

6
Fuel to the Flame

Since the beginning of recorded history man has used his abilities as a toolmaker to try to ensure for himself a continued rise in his standard of living and in the number of his possessions. The rise has been a steady one, interrupted only by periods of war, plague or famine. Communities have accepted and used inventions which would contribute to this pattern of development, and rejected those which would not. A hundred and fifty years ago, with the emergence of the American System of Manufacture, the rise in the material standard of living accelerated dramatically. Although originating in the cotton and armaments industries, the concept of producing finished goods from raw materials under one roof, using a centralized power supply, lent itself readily to many other uses. The production line has in the last hundred years radically altered the rate at which innovation affects our lives. One of the results of that change has been to narrow the gap, at least in the developed world, between rich and poor, as the profusion of goods available on the market has helped to create a certain democracy of possessions. Another has been to bring into being organizations whose aim is to stimulate a desire for novelty, through advertizing.

A new product is rarely if ever aimed at answering a basic need to survive, but rather to appeal to the desire for 'change for change sake'. Thus a circular pattern has emerged in which the creation of demand is linked to the rate of growth of the economy, which in turn is linked to output levels, which are linked to employment, which depends for its continued existence on the stimulation on consumer demand. As a result, the production line dominates every aspect of our

The frozen centuries, illustrated in a Flemish painting of 1585. The years of short summers and severe winters, between the thirteenth and fifteenth centuries, spurred the development of many inventions which were to change the style of European domestic life.

daily lives. Masses of people move to and from work at the same time every day, and so transport systems are tailored to meet these peaks in demand and are underutilized for the rest of the day. City planning concerns itself with the provision of housing whose position is dictated by ease of access to industrial areas. The making of breakfast at the same time by millions of people across the country is as much a part of the production line as is work at the conveyer belt an hour later. Our behaviour has been profoundly altered by the production line. Standardization increases efficiency of output, and as a result, only the rich can afford possessions that are truly unique. The rest of the community lives in the same kind of house, wears the same pattern of clothes, drives the same car, watches the same cable shows, dreams the same dreams of success in the hierarchy of work. The trend towards providing goods for everybody at prices they can afford – the laudable ambition of the pioneers of the American System – has led to economies of scale. Mass production is cheaper than short runs. The supermarket has been replaced by the shopping mall, the conglomerate by the transnational corporation.

As the sources of supply become more and more centralized and we face the inevitable shortages of raw materials and energy, the question is to what extent our society has become inflexible. How will we adapt to the necessary change in our life style? How can one light a fire in a building with no fireplaces? What will happen as, with dwindling energy supplies, life becomes colder? The last time a significant change of this nature happened was in early medieval Europe, when, as now, the fire in the house was everybody's fire.

The mid-twentieth-century artefact par excellence, *the automobile – the most ubiquitous example of reliance on a single source of energy.*

This Los Angeles freeway interchange is a typical response by modern urban planners to the demands placed on them by the production line and the mass behaviour it dictates.

Towards the end of the Dark Ages the manorial system had evolved over much of Europe. It originated from the need for groups of peasants to live and work together in order to pool their resources. The number of oxen needed to pull the new heavy-wheeled plough led to the sharing of plough and ox, and then to housing and maintaining these in a common service unit, and thus, in time, to living round that unit. As a reflection of the uncertainty of the times, the manor became self-sufficient. In certain areas it was against the law for the manor to use any object or commodity which could not be produced within its boundaries. More has been recorded about life in the Early English manor than in any others, which serves as a useful starting point since it was in England that the major changes round the hearth began. Saxon society was rigidly hierarchical. A person's position was determined by his worth, murdered: the penalty for murder was either death, or payment of the *wergild*, or body-price, of the victim. A noble was worth 1200 shillings; a thane, 300; a churl, or free peasant, 200. Serfs were worth nothing, and a man of whatever rank was worth more than a woman of equal social standing.

Life on the Saxon manor, even in times of peace, was for the peasants nasty, brutish and short. Most of them worked dawn to dusk from the age of six until death, usually before the age of forty. For the churls, thanes and lords of the manor, life was a little better, and it was centred round the fire. Again and again the poets of the time speak of the warmth and companionship of the fire in the hall. Comforts there were few. Household utensils such as spoons, bowls, cups and plates were made of turned wood. Furniture consisted of rough-hewn tables, benches and stools. Only the leading members of the group would warrant a separate chair. Leather was in evidence everywhere, as slippers, shoes, gaiters, bottles, reins, trappings for the animals, halters, bags and purses. Clothes were predominantly made of a rather coarse wool, ranging in quality from a fine flannel-like material to something approaching the texture of Harris tweed. Linen was used for underwear. A well-dressed member of the royal family would wear linen long johns underneath a pair of wool trousers, and a three-quarter length woollen cloak, belted at the waist and trimmed with silk. In winter he would wear a short fur cape as well, and over that, a long woollen mantle held at the shoulder with a brooch. Throughout the year a peasant would wear a rough wool tunic and trousers, and no underwear. Cloth was woven on the manor by the women, whose main occupation—together with baking—it was. The diet was rough: coarse bread, gruel, cheese, vegetables, eggs, boiled mutton and bacon, an occasional chicken, a rabbit, all washed down with ale or mead. To judge by their skeletons, people seldom grew above 5′ 6″ in height. The women suffered from a type of arthritis, caused by spending a lot of their time squatting.

Their houses, or halls, were equally uncomfortable. The larger ones were traditionally rectangular in layout, and there was rarely an upper floor. The houses were almost entirely constructed in timber, itself a word relating to the Old English verb 'to build'. The raw wood was occasionally, in the richer houses, adorned with tapestries to add warmth and colour. These hangings were often mentioned specifically in wills, one of which said: 'I grant to St Peter's monastery at Bath . . . a set of bed clothes with tapestry and curtain . . . and to Eadgifu, two chests, inside them the best bed-curtain and a linen covering and all the bed clothes that go with it.' The main room in the house, the great hall, was the principal room for eating and sleeping. During the day rushes strewn on the floor absorbed the spilt liquids and collected the debris of meals, and in the spring and summer months fragrant flowers and herbs would often be added to clear the atmosphere. At night the rushes were replaced with fresh ones, and the company slept either on these or on flock mattresses placed on the floor. If all this sounds painfully chilly to the modern reader, it should be remembered that the average temperatures of northern Europe were several degrees higher than they are today. Apart from making living more comfortable this also provided a growing season at least three weeks longer than it would be in the present century. This medieval Indian summer was not to last, however, and when it changed a great deal more than the temperature was to alter.

Among the earliest references to the change comes from the *Anglo-Saxon Chronicle*, kept by monks, for the year 1046: 'And in this same year after the 2nd of February came the severe winter with frost and snow, and with all kinds of bad weather, so that there was no man alive who could remember so severe a winter as that, both through mortality of men and disease of cattle; both birds and fishes perished through the great cold and hunger.' Ice cores taken up by modern scientists from the ice-caps of Greenland and the Antarctic, containing bubbles of trapped air, vegetation and organisms from the period, indicate that throughout the early part of the twelfth century the climatic change really began to make itself felt, with increasingly uncertain weather, heavy rains, gales and lower summer temperatures. At the start of the thirteenth century what is now called the 'Little Ice Age' had set in. It was to last for nearly two hundred years. After centuries of relative warmth, this sudden onset of cold triggered widespread changes in the living patterns of the European community.

The chief stimulus to change was the need to stay alive through winters that became increasingly severe, as the monks had noted. The first innovation to come to the aid of the shivering communities was the chimney. Up to this time there had been but one central hearth, in the hall during winter, and outside in summer. The smoke from the central fire simply went up and out through a hole in the roof. After the weather changed this was evidently too inefficient a way of heating a room full of people who until then would have slept the night together. The lord and lady may have pulled a curtain round themselves, but everybody shared the warmth of the one central fire.

The forerunner of the chimney that was to answer the problem – and cause such widespread change – may have been similar to the system used at the monastery of St Gall in Switzerland, in the ninth century. There, underfloor heating was vented through chimney stacks built separately from the building. Although the primary purpose of the stacks was to conduct the fumes away, it may have given someone the idea of adapting the stack to the very simple type of blast (draught) furnace in use at the time. The understanding of the physics of updraught and downdraught in the use of chimney breastwork and hood, which are what enables the fire to both keep going and give out heat and at the same time carry away smoke and sparks, indicated experience of the use of draught power. The only

A painting (1438) of St Barbara, incorporating the latest luxuries afforded by the presence of the chimney. Wooden furniture keeps her feet off the tiled floor, glass windows are backed by storm shutters, and beyond the chimney-hood a towel hangs above a ewer full of warm water. The chimney stimulated indoor activities such as reading.

other furnaces around at the time were those using charcoal for making iron and glass, and the expertise very probably came from the craftsmen who used these.

The building to which the new chimney was added had already begun to change in reaction to the bad weather. The open patio-style structure had been replaced by a closed-off building, built to withstand violent meteorological changes. The new chimney, whose earliest English example is at Conisborough Keep in Yorkshire (1185), also produced structural changes in the house. The use of a flue to conduct away sparks meant that the centre of the room was no longer the only safe place for the fire. To begin with, buildings were by now less fully timbered so the risk was less, and the flue permitted the setting of the fire in a corner or against a wall. The extra insulation needed for safety introduced more brickwork (a good fire retardant), and as the section of wall by the chimney was thus strengthened, the chimney began to act as a spine that could be used to support more than one room – which, in any case, could now be heated by separate fireplaces. The hood on the fireplace prevented sparks from reaching the ceiling, and as a smaller room could more readily be heated than a larger one, the ceilings could now be lower.

The primary effect of the introduction of these new rooms was to separate the social classes. The first chimneys in royal residences were constructed in rooms to which the king could withdraw, at first with his immediate family, later with officials. The English Privy Council did not come into existence (in 1300) before there was a place to be private. Special state apartments were constructed, and separately heated. Bathing became more common. The smaller rooms also meant that paperwork could continue through the winter. (Again and again, from the fifth to the eleventh century, we read of monks complaining that their hands are too cold in winter, and that sometimes the ink freezes in the pots.) Separate accounting offices were built, like the one in Abingdon Abbey, in 1260. As the general economic pace of Europe was quickening at exactly the same time, due to the innovative use of water power on an extensive scale, there was more paperwork than there had been previously, and so the ability to work through the winter as a result of the introduction of the chimney was of fundamental importance in the commercial recovery of the continent.

The chimney also introduced the concept of privacy for the first time. The new heated bedrooms, where people slept away from the general community, and often naked, altered as well the attitude to love. It now became a personal, private, romantic activity. Courtly love-poetry may first have been written during long periods of abstinence on the Crusades, but it would not have flourished in the cold of northern Europe without some help from the chimney. Interior decoration too

The new indoor hygiene. A bath-house scene from Germany (c. 1470) shows the kind of private activities which led to severe censure from the Church, as the new warmth encouraged intimacy and 'licentious behaviour'.

changed to fit the altered climate. Plaster first came into use round the chimney hood as an efficient sealant against draughts, and soon spread round the rest of the room. In time, the plaster was moulded into soft lines and shapes, and painted with bright colours. Initially these painted walls were made to look like brickwork; later they became the vogue in their own right, in rich warm shades of golds and yellows. People took to painting their coats of arms on the walls and above the fireplace. In Winchester, in 1239, a map of the world was planned for the Great Hall. This decorative habit soon spread down to the middle classes who began to adorn their homes with scenes of hunting. By the thirteenth century the coloured plaster was giving way to wainscotting and panelling as a more efficient anti-draught system. The wood used was often fir, easier to work than oak, but foreign. In 1252 Henry III ordered this wood from Norway, 'to wainscote therewith the chamber of our beloved son Edward, in our castle of Winchester'. Woollen hangings were still used to add warmth, as they had done in pre-chimney days. The simple cloth hangings originally used in the hall behind the high table, to protect the family and their guests from draughts at their backs, with increasing wealth had become rich tapestries. These were made chiefly in

Right: The Buxtehude altarpiece. This is the earliest illustration (c. 1395) of knitting.

Below: John of Portugal entertains John of Gaunt, on his right, in a late fifteenth-century dining chamber. Note the tapestry hung to keep out the draught from the unlit chimney, the table napkins, and the hatchway to the separate kitchen.

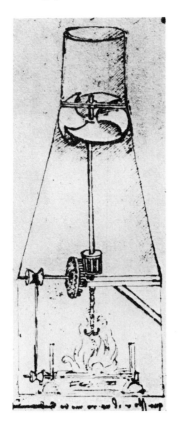

A design (c. 1480) by Leonardo da Vinci for a turbine-operated roasting spit.

the French town of Arras, from which all tapestries took their name. By the fifteenth century they were in use in every room.

This clothing of the house against cold was matched by changes in personal clothing. Two major innovations occurred by the fourteenth century, at the latest: knitting, and the button. The earliest buttons are to be seen on the Adamspforte in Bamberg cathedral and on a relief at Bassenheim, both in Germany, near Hamburg, around 1235. The first example of knitting is depicted on the altarpiece at Buxtehude, where the Virgin Mary is shown knitting clothes for the infant Jesus. Both buttons and knitting contributed to closer-fitting clothes that were better at retaining heat. The effect on mortality can be seen in contemporary changes to what was said in wills: children were expected to survive, and provision was made for them. As the houses became warmer, the life style within them could become more refined. Lead conduits began to carry water to washbasins and privies in apartment rooms. In the thirteenth-century castle of Frederick II, at Castel del Monte, the privies flushed. This improved hygiene led in turn to better manners at table. Pages are shown holding jugs and linen towels for the diner to clean his hands on. Stroking dogs under the table was now frowned upon. The kitchen, of course, was now a separate room, and had been so from soon after the introduction of the chimney. Some of the bigger houses had kitchens with multiple fireplaces, served by a common chimney. The understanding of draught physics may have been improved by the arrival of Tartar slaves into Italy in the fifteenth century, bringing with them air turbines with which to power fireplace spits.

By that time, the separation of the classes was complete; the ties between them that had been expressed in the act of sleeping together at a common fire each night were broken. The tightly knit, agriculturally based feudal world had gone up the chimney. Not everyone approved. The poet William Langland had written towards the end of the fourteenth century:

> Woe is in the hall in all times and seasons
> Where neither lord nor lady likes to linger
> Now each rich man has a rule to eat in secret
> In a private parlour, for poor folk's comfort;
> In a chamber with a chimney, perhaps,
>   and leave the chief assembly
> Which was made for men to have meat and meals in.

By the 1540s England was enjoying a rapidly rising standard of living due to the general recovery of Europe from the Black Death, the increasing wealth from stability at home and aggressive expansion abroad, and above all to the wealth soon to be released by the final effects of Henry's dissolution of the Catholic monasteries. The economy was a seller's market, and this expressed itself in fixed rents and increasing prices for commodities, all of which tended to favour the middle class and the agricultural yeomanry. These people had by now absolute tenure of their properties, and rents that had been fixed centuries before. Many of the rents were at peppercorn level. By this time the middle classes were beginning to enjoy a cash surplus, which they spent on the luxuries that previously only the very rich had been able to afford. Much of their spare cash went into building: either new houses, or refurbishment of medieval ones. The chimney and the innovations that followed it permitted the husbandman, just as much as the king, to erect sufficient rooms to house his family, the size of which was increasing due to a better diet and the absence of war.

One of the finest examples of the building boom – and of the reason for the changes it caused – is to be seen at Hardwick Hall, still standing today in the English county of Derbyshire. It was built between 1591 and 1597 by Bess of Hardwick, Dowager Countess of Shrewsbury, who moved into the new house at the age of seventy-seven. She was a parvenue, a formidable character who had been married four times, outlived all four husbands, and, as the verse went at the time:

> Four times the nuptial bed she warmed
> And every time so well performed
> That when death spoiled each husband's billing
> He left the widow every shilling.

Hardwick Hall: the west front. The initials E. S. (Elizabeth Shrewsbury) can be seen surmounting the towers. Bess maintained her own glass-makers in the grounds of the house while it was being built.

Hardwick Hall: the High Great Chamber. Note the wainscotting, wooden panelling, plasterwork and tapestries for keeping out the draught. Woven rush-matting covers the stone floor.

Hardwick Hall represents more than an old woman's expensive foibles, however; it was known, for its extraordinary use of the material, as 'Hardwick Hall, more glass than wall'. It was this prolific use of glass in the new buildings in England that was to have far-reaching effects. Although there had been an active glass industry before this time it had in the main been confined to cathedrals and palaces, and as the surge of cathedral building waned in the fifteenth century, so had the industry. But with the growth of building in the sixteenth century demand for glass so far outran production that foreign glass-makers were attracted to the opportunities afforded by the English market. By 1567 much of the glass in use in England was being imported from the Continent. In the same year permission was given to a certain Jean Carré, a native of Arras who was living in Antwerp at the time, to enter England and set up furnaces for making window glass in the forest of the Weald of Kent, to the south-east of London. One of the conditions of entry was that he teach his art to the English, so that on the expiry of his patent after twenty-one years there would be a thriving native industry. As it happened, Carré was to die well before the term expired, but by 1580 there were many more foreign glass-makers moving around the forests of England. They had become itinerant because of the rising price of wood, and of transport. Wood was only cheap if it was bought close to the furnace, and once the immediate area had been cleared, it paid to move on closer to fresh supplies.

Several things conspired against the glass-makers. They themselves were stripping the woods to service the building boom that had brought them to England in the first place. So were the iron-makers, burning wood, as the glass-makers were, to make charcoal for their furnaces. In 1581 Parliament enacted legislation forbidding the felling of trees within twenty-two miles of the Thames, within four miles of the great Sussex forests, and within three miles of any part of the coastline. The increasing number of glass-makers now started to feel the effects of high demand and limited raw materials. Whenever possible, they had settled near rivers on which they could transport their raw materials and the finished product more economically than over land. So in 1593 a bill went before Parliament seeking to outlaw glass-making furnaces within eight miles of any river in the country.

The glass that was then so much in demand came in two types, according to whether it was needed for small lattice windows or larger frames. The latter used broad glass; this was blown into a cylinder when molten, then the glass cylinder was split down one side and, still hot, rolled flat. The other type of glass was crown, and was made by blowing a bubble of molten glass on the end of a blowpipe. The bubble was then fixed to the end of an iron rod, the blowpipe connection was cut, and the rod was spun so that the bubble widened and flattened under the effect of centrifugal force. The flat, circular plate was then broken off the end of the rod. This kind of glass can still be seen in older houses, where it is identifiable by the familiar bull's-eye at its centre where it was connected to the rod.

So, by 1600, England was facing an acute timber crisis, thanks largely to the increase in glass production. Attempts were made to solve the problem by looking to overseas supplies of raw materials, and a glass-making furnace was even set up in Jamestown, Virginia, in 1608, only a year after the colony had been founded. (One of the men on the council that invested the colony with its charter was a Sir Robert Mansell, who within a few years was once more to play a leading role in events.) In spite of the fact that Jamestown had plentiful local supplies of sand, potash and wood, needing to import only lime, the business failed after a year, principally due to lack of skilled labour. The price of glass continued to rise in England, and the wood famine became worse. Partners in crime with the glass-makers were the ship-builders, themselves taking advantage of the economic success of the country, and the cannon-founders whose guns were in demand aboard the ships. The legislation earlier aimed at reducing the wood-cutting activities of the glass-makers had been in part an attempt to preserve wood for ships and guns. The drive to make England self-sufficient in terms of defence, and to return to the great days when she ruled the wool market, was to put further strains on the wood supply.

Sixteenth-century glass-makers at work. The domed furnace is lined with brick to retain heat. In the foreground (E) lies a mould. In front of the furnace, blowers begin to shape the molten glass blob prior to blowing, while outside a porter carries away finished articles packed in straw.

The iron-founders of the south-eastern forests had been casting iron cannon since 1543, when a clergyman called William Levett had developed the technique. These cannon, however, being subject to royal monopoly, were being exported in such numbers (to bring in revenue) that by the latter end of the century Sir Francis Drake was complaining to Parliament: 'I am sure heretofore one ship of her Majesty's was able to beat ten Spaniards, but now, by reason of our own ordnance, we are hardly matcht one to one.' In spite of the popularity of iron cannon abroad, bronze was considered a better material by the English, and when the Spanish blocked Antwerp in 1564 supplies of copper from which the bronze was made dried up. Under Elizabeth, the Secretary of State Lord Cecil was anxious to establish a copper industry in England. Little copper had been mined here, although as all mining of precious metals was a monopoly of the crown, Cecil was in a position to take whatever steps necessary since copper was often found in the same deposits as gold and silver. So he had the mines, but lacked the expertise. For this he turned to the

Mining operations in 1544. The mine is sited by a river where a barge (right) awaits loading with ore and metal. Above, left, wagon loads of ore are brought down to the water-powered blast furnace. Note the aqueduct bringing water from the distant mountain, and wooden trolleys running on wooden rails at the mouth of the tunnel behind a man with a wheelbarrow.

Germans, and in particular to a family from which the kings of England had been borrowing money for a hundred years. These were the Fuggers, directors of a great European banking house established in Augsburg, Bavaria. It seems they had close connections with another financial concern run by a man called David Haug, who had control of mining operations in the Tyrol, near Innsbruck, where much copper was already being produced. In 1564 Cecil invited the Germans to come prospecting, and three prominent members of the Haug firm arrived in England. By 1565 they had found copper in Cumberland, and proceeded to bring in nearly four hundred Tyrolean miners to mine the deposits. Work went ahead in spite of trouble with the local people, which at one point caused Elizabeth to write to ask local Justices 'to repress the assaults, murders, and outrages on the Alamain [German] miners lately come there for the purpose of searching'. In the summer of 1566 a rich copper mine was found, and financial support began to increase. Two years later Cecil ensured control over events by incorporating a company known as the 'Mines Royal'.

Parallel with the copper-mining activity went Cecil's other interest, wool. Since the fourteenth century this industry had been the backbone of the English economy; however, part of the processing involved combing the wool in such a way as to matt the fibres together, and for this metal combs were used, principally made in brass. Again, there was no brass industry to speak of in England, and Cecil wished to establish one. The making of brass – an alloy of copper and zinc – involved the use of another ore called calamine, which had to be imported. Once more England turned to the Germans. This time the initiative came from the Master of the Mint, who helped a certain Christopher Shutz from Saxony to get a patent on a method for beating brass ingots into flat plates, known as the battery method because the metal was battered by trip hammers. Together with this patent came authorization in 1565 to go on to any land anywhere in the country to prospect for calamine. In June of the following year the foreign miners found calamine, in the Mendip hills close to the coast. Within a few months they had extracted twenty tons, an event considered of such national importance that only Cecil and the Master of the Mint knew of it. So in 1568 another company was incorporated, the Society of Mineral and Battery Works. They and the Mines Royal had between them monopoly over copper and brass production in England, and the investors, sensing the profits, urged them on. Production increased, and so did the need for charcoal. The trees fell faster. By 1615 the timber crisis was acute. The answer to the problem already existed, and was to be one of the instruments of profound change not only in England but throughout Europe in the century that followed: it was a new type of furnace.

In 1611 a group of people, led by Edward Zouch, a courtier with an eye to money, was granted a patent for a new way of making glass without the use of wood. The fuel in the furnace, probably designed by a glass-worker called Thomas Percival, was coal. Most of the coal in use at that time was mined at the surface, and was therefore full of impurities. The new furnace solved the problem of the effects on the glass of impurities in the fuel by using an underground tunnel to the outside of the furnace, to draw in air. The incoming air entered the hearth at the centre of the furnace, beneath a circular grate. Round this grate were placed the pots containing the glass mixture, covered with lids to protect the glass from soot droppings. When the coal was fired on the grate, the special clay lining of the dome-shaped oven roof reflected the heat back on to the pots, raising the temperature in the oven, and the hot air and fumes escaped through flues along the wall. The reverberatory furnace, as it was called from this reflecting function, produced glass that the examiners compared favourably with glass made in the traditional wood-burning furnaces.

That was enough for Sir Robert Mansell, the man who had been involved originally with the ill-fated Jamestown venture. He bought his way into the Zouch company. Mansell was a powerful figure, a courtier and a vice-admiral of England, whose family had considerable coal deposits on their land. In 1615 Mansell offered to buy out his partners and pay off Zouch's debts; in the same year James I issued a proclamation which forbade the use of wood in glass-making. The Proclamation Touching Glass stated that 'no person or persons whatsoever, shal melt, make, or cause to be melted, any kinde, forme or fashion of Glasse or Glasses whatsoever, with Timber, or wood . . .' Mansell now had a monopoly, and within a few years he had begun working the first major coalfields in Britain, in Northumberland and Durham. Then, in 1622, Mansell went before the Privy Council to defend himself against a case brought against him by furnace-makers. On the committee hearing the case, which Mansell won, was another courtier, the Viscount Grandison, whose family was soon to become involved in the new furnace.

That step was aided by the advent of the Civil War in 1650, when the two royal monopoly societies were suspended. They had been in financial trouble since 1640, and were criticized for not being able to meet domestic demand for copper and brass. After the Restoration attempts were made to re-establish the societies, but in those ten years they had lost ground to private enterprise, and they were eventually disestablished in 1689. Meanwhile the grandson of the Viscount who had listened so keenly to Mansell's case had now, in 1678, obtained a patent for using the same reverberatory furnace for smelting lead, using coal as the fuel.

Grandison chose a site near Bristol, close to the lead deposits in the Mendip hills. With the earlier discovery of calamine in Somerset, and the finding of copper deposits in the Forest of Dean, Bristol was rapidly becoming a centre for metal-working. It was ideally suited as such, with raw materials close at hand, coal available nearby, the river Avon as a power source, and good port facilities. The lead works was set up, and a certain Clement Clerke was hired to run it. But a dispute arose concerning Clerke's salary, unpaid for several years, and receiving no satisfaction from Grandison, Clerke left, taking his assistants with him. Turning to copper, Clerke now looked for a way to adapt the reverberatory furnace to smelt it. The problem lay in the higher temperature which is necessary for working copper. One of Clerke's assistants, John Coster, solved the problem, using either a more efficient draught system or less impure coal. By 1691 copper mining had developed to such an extent that Coster had left the Clerke organization and was managing another company. In the decade that followed the copper smelters began to look for new ways of disposing of their stocks, other than for armaments and money.

The real opportunity lay in brass, which, in spite of Lord Cecil, was still being imported in large quantities. After the restoration of the monarchy England's economy had surged, and the demand for household brass rose rapidly. This may have been incentive enough for a young Quaker from Birmingham, Abraham Darby, to move to Bristol with the intention of making brass – and profit, although his move may have been facilitated by the copper smelters who saw a chance to get rid of stocks.

Bristol was attractive to him not only commercially, but because it was the second biggest centre of Quaker worship in the country. Darby had been apprenticed in Birmingham to a Quaker firm engaged in making brass malt-mills, small machines somewhat like modern hand-operated coffee grinders, which were used for crushing malt in domestic brewing. He owed his expertise in brass to his religion: no Quaker would work in armaments, where most of the iron and copper went. So Darby came to Bristol with the qualifications Bristol wanted. The only other raw material he would need for his brass was calamine, and there was plenty of this within fifteen miles of the city.

Darby's first move was to seek help from those with experience in brass-working on a scale hitherto unknown in England: the Dutch. In 1704 he went to Holland to recruit for his Bristol Brass Wire Company. He was keen to learn the new Dutch technique for beating brass with automatic trip hammers, a facility which made their brass hollow-ware the best in Europe. Dutch workers were soon on their way to Bristol, and over the next few years Darby was to make more visits to Holland. During one of these later journeys he observed the

Dutch casting brass pots in sand, and was struck with the possibility of casting iron by the same method, since iron being very much cheaper than brass would sell more easily on the domestic market. The Dutch had already tried, and failed, with iron. One of Darby's young apprentices, an otherwise unknown man called John Thomas, came up with a method in 1707, and Darby was so sure of its importance that when he bound Thomas to him he stopped up the keyhole of the door to the room in which they came to their legal agreement. John Thomas had given England the key to the Industrial Revolution.

Within a year Darby had left Bristol and brass-making for good. He had moved up river to Coalbrookdale, where he had heard there were plentiful coal deposits for his furnaces. By one of those strange quirks of history, in the same year that Darby had moved to Bristol and begun an independent career, less than a hundred miles from him another man had done something which set in motion a chain of events that would also end in Coalbrookdale. His name was Thomas Savery, and he had built a drainage pump. Drainage had become a problem throughout Europe as mining operations increased. In

Right: Savery's pump, known as the 'Miner's Friend'. Once steam was forced down the tube to water level, a valve let the water in, the steam condensed, and the partial vacuum thus formed sucked the water up the tube.

Below: The Coalbrookdale iron works by night. At the beginning of the eighteenth century most industry was on a small scale, and more often than not set in rural surroundings close to wood, coal and water supplies.

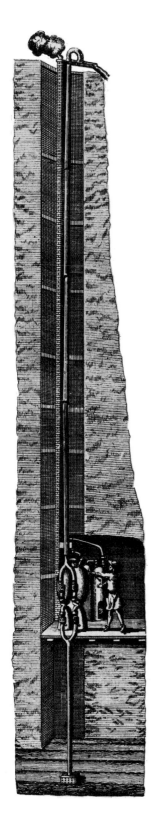

Cornwall, where Savery worked, the problem was acute, since many of the mines there were sunk along the coastline and extended out under the sea. As early as 1695 Celia Fiennes, the indefatigable lady traveller, had written of her tour of the West Country: 'They even work on Lord's day to keep the mines drained – one thousand men and boys working on drainage of twenty mines.' At Redruth she saw 'a hundred mines, some of which were at work, others that were lost by waters overwhelming them'. The problem of lifting water by pump had led shortly after the beginning of the seventeenth century to the discovery of the vacuum, and in turn to the vacuum pump developed by Otto von Guericke in 1654. Guericke's pumps had stimulated further experiment, such as that in 1695 by the Frenchman Denis Papin, who had tried to operate a piston pump with the force of gunpowder, and later with steam power. His steam pump used the pressure of steam to force the piston up the cylinder, but he had to use weights to lower it again. Savery took this idea further with his pump, nicknamed the Miner's Friend. He abandoned the use of a piston, and forced steam into a 'receiver', which was part of a tube extending downwards into the water to be drained. The pump failed because the quality of the metalwork could not withstand the high steam pressure in the first half of the pump cycle; the soldering would melt under the heat, and the joints would split. Strangely, it was this very type of failure that was to bring success to another man who was working on the same problem. His name was Thomas Newcomen, an ironmonger living in Dartmouth.

Newcomen and his partner John Calley, a plumber, travelled round from mine to mine dealing in metal or doing work on site, so they were well acquainted with the importance of drainage. At some point between 1700 and 1705 they produced their 'engine', using ideas from the work of Guericke, Papin, and above all Savery, whose use of a separate boiler to generate steam was crucial. Newcomen, however, revived the idea of the piston. The giant pumping engine received its steam from a beehive-shaped boiler which was at the time already in use in the brewing trade for boiling up malt. Like the brewers, Newcomen sited his boiler in a brick housing directly over the grate, where the hot air circulated over the boiler before leaving through a chimney. He sealed off the top of the piston cylinder by a layer of water lying on top of the piston itself. The process was slow – too slow.

But then, in 1704 or 1705, the accident that Savery had encountered befell Newcomen also. His solder melted. In his own words:

> the cold water that had been allowed to flow into the leaden jacket, surrounding the cylinder, penetrated the cylinder wall through a casting fault mended with tin-solder, which had been melted by the steam. Forcing itself into the cylinder, the cold water immediately condensed the steam, causing such a high vacuum that the weight hooked onto the one end of the little beam, which was supposed to represent the weight of water in the pumps, was so insufficient, and the air pressed with such tremendous force on the piston its chain broke and the piston itself knocked the bottom of the cylinder out and smashed the lid of the small boiler. The hot water flowing everywhere convinced also their senses, that they had discovered an incomparably powerful force . . .

They undoubtedly had, and within weeks Newcomen had redesigned the engine so that his 'accident' happened once every stroke of the piston. He added a mechanism to introduce a small jet of cold water into the piston just as it reached a full head of steam. The Newcomen engine was an instant success, and within ten years it was in operation all over Europe. The only problem was that as users began to demand bigger and bigger cylinders expense became a critical matter, since brass was used to make the cylinders, and brass cost a great deal. In 1722 Newcomen found the answer to his need, in Coalbrookdale.

In the intervening years Darby had drawn on his experience with the brewers to raise the temperature in his furnace to the point where it would both melt iron for casting, and leave it untainted by impurities. It happened that the coal in the area was particularly free from sulphur, so it was relatively pure. Darby cleaned it further by coking it before use, as the brewers had been doing for some years in their malt-

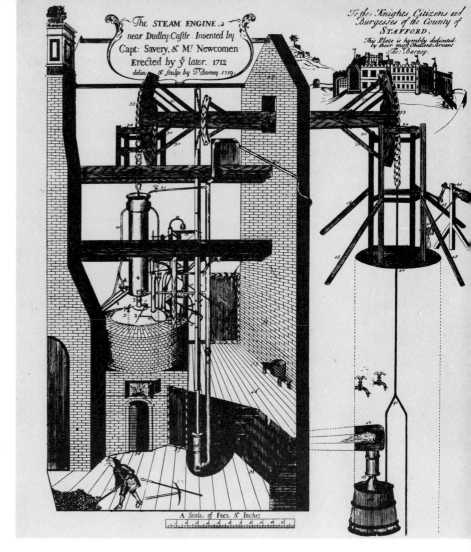

The STEAM ENGINE
near Dudley-Castle. Invented by
Capt: Savery, & Mr. Newcomen
Erected by ye later. 1712
delin: & Sculp: by T. Barney. 1719.

To the Knights, Citizens and
Burgesses of the County of
STAFFORD.
This Plate is humbly dedicated
by their most Obedient Servant
Tho: Barney.

A Scale of Feet & Inches

*Newcomen's pumping engine.
Steam entered the cylinder above
the circular brick boiler-house
and as it did so was condensed by
a jet of cold water. The vacuum
formed by the condensation
caused air pressure to push the
cylinder piston down. This piston
was connected by a chain to one
end of a beam, pivoted so that as
the cylinder end dipped the other
end rose, lifting with it the
suction pump rod (right).*

drying operations. When Thomas Newcomen came looking for large
iron cylinders to replace his expensive brass ones, he found Darby's
foundry producing molten iron in sufficient quantity to cast them, at a
price that was very much less than that of brass. From then on the
Newcomen engine never looked back – until one day, forty-one
years later, in Glasgow University. What happened there was to turn
the pump into a machine that drove the Industrial Revolution.

The myth of the events at Glasgow is still taught in schools: that
James Watt was watching a steam kettle boil one day, and noticed the
force of the steam lifting the lid, and because of this observation was
able to invent the steam engine. In fact he was then working as
instrument maker and repair man to the university, and he was asked to
fix a model of the engine. While doing so, it occurred to him that it was
inefficient because while the jet of water condensed the steam in the
cylinder very quickly, it also cooled the cylinder down, causing pre-
mature condensation on the next stroke. He realized that, in effect, the
cylinder had to do two things at once: it had to be boiling hot in order

to keep the steam from condensing too early, and it had to be cold to
condense the steam at the right time. The cylinder could not do both,
and Watt saw that a different system was needed. He designed a
separate condensing unit, which meant he could keep the cylinder hot
by jacketing it in hot water from the boiler. This, in its final form, was
the Watt steam engine: immensely more powerful than the Newcomen engine, and capable of being used as a source of power.

*Watt's pumping engine,
showing the separate condenser
to the left below the main
cylinder. The vacuum pump is
the cylinder to the left of the
condenser, which is immersed in
water to keep it cool.*

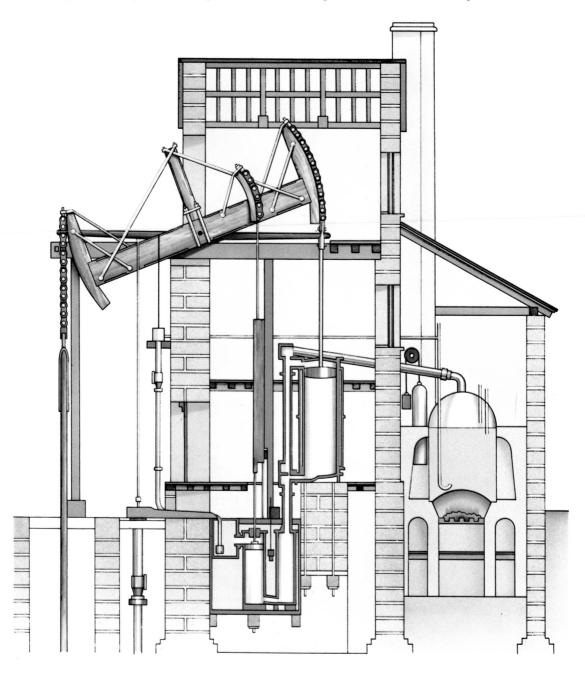

The very success of the design faced Watt with a problem. Because to have retained Newcomen's water seal on top of the piston would have cooled the cylinder down, Watt dispensed with it. Now the piston had to be a perfect fit, so that every ounce of steam power could be used. When Watt's patent was granted, in 1769, no one could cast cylinders accurately enough to achieve that precision. The answer came in 1773, thanks to the activity of a commission set up in France to inquire why cannons were blowing up in the faces of their gunners so often that the French feared their own cannon more than those of the enemy. That year a French brigadier called de la Houlière arrived in England to learn about new methods for making safer cannons. During his visit he met two brothers, John and William Wilkinson, who were operating an iron foundry near Coalbrookdale. The meeting was to forge a bond with France that would take William to Nantes after the Revolution as manager of the new state foundry there. The other result of de la Houlière's visit was John's development of a cannon borer, a year later.

His system for making cannon was entirely new. He first cast the guns in a solid piece, then mounted the work horizontally, turned it with a Watt steam engine, and as it turned, bored it with a cutting head fed slowly and accurately down the centre of the barrel by means of a gearing system attached to a guide bar. This in itself was an extraordinary innovation, introducing as it did for the first time the principle of the guide in machine tools. By 1775 Wilkinson had a machine that would bore a cylinder, 'so that it doth not err the thickness of an old shilling'. This was precise enough for Watt.

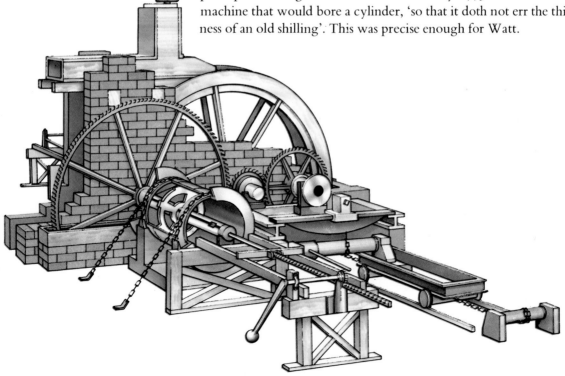

Wilkinson's cylinder-boring machine. In this version the power source is an overshot waterwheel behind the brick wall. The secret of the accuracy achieved by Wilkinson lay in his supporting the shaft of the cutting head so that as it rotated there was no deviation off axis.

Watt was not the only man for whom Wilkinson provided gun barrels. He also provided them for his brother-in-law, Joseph Priestley. In 1762 Priestley had married John's sister Mary, then at the tender and impressionable age of eighteen. Mary's new husband was a preacher who failed in the pulpit because of his stammer. He was, in common with the times, a scientific dabbler – and, to judge by his letters, he was also permanently afflicted by concern for money. He had evidently married Mary for her money, and when he found out that she had little, had turned to her brothers. By the late 1760s he and Mary, supported by John, were living in Leeds, where Joseph divided his time between preaching badly enough to be fired, and investigating the behaviour of gases in the brewery next door. He was particularly interested in the layer of 'fixed air' above the beer in the vats, and observed that it was 'generally in a layer about nine inches or a foot in depth within which any kind of substance may be conveniently placed'. The things he placed in it were many and varied, including mice (which died), candles (which went out), small charges of gunpowder (which were muffled), and quantities of ether. Unfortunately this last was to cause his downfall, because before he could observe what happened to it he let it fall into the brew, at which point the brewers threw him out. He had, however, noticed one other thing. If he poured water from one glass into another within the layer of 'fixed air' above the vats, the water became impregnated with bubbles. The fizzy mixture had a pleasant taste, he noted. Having failed to get Admiralty interest in the new water as a cure for scurvy, he lowered his sights. Soda water, as it was called, rapidly became the rage of health spas throughout Europe, and Priestley's name was made.

A typical eighteenth-century brewery. The 'fixed air' Priestley observed was given off in frothy bubbles of carbon dioxide as a result of the fermentation process.

Fig. 1.

Priestley's apparatus for experimenting on gases (c. 1775). As a result of his work on the constituents of air Priestley discovered the existence of oxygen, and spurred interest in respiratory medicine.

His preoccupation with the behaviour of gases next led him to use his brother-in-law's free supply of gun barrels. He used them as containers for various substances which he burned, collecting the gases that came out of the barrel as a result and submitting them to a battery of tests. One such test was designed to check the purity of air. Two wires were introduced into a glass vessel, each connected at the other end to an electrostatic generator. Inside the glass, the wires were brought almost to within touching distance; then the gas to be tested was pumped in, and a charge sent down the wires, causing a spark. If the air ignited, said Priestley, it was pure. The device became known as a eudiometer, and it aroused considerable interest in Europe among both scientists and those interested in hygiene. Priestley wrote to several people about his work, including an Italian savant called Alessandro Volta, who was living in Como at the time.

Volta received the letter just after he had invented his own portable electricity-maker, the electrophore. This consisted of a cake made from three parts turpentine and one part wax, which, when rubbed, would generate static electricity; it would retain the charge almost indefinitely, for which reason Volta called it his Eternal Electricity Machine. As soon as Volta heard about Priestley's eudiometer, he had himself one made in the form of a glass pistol, with two wires charged by his electrophore. In November 1776 Volta discovered, during a fishing trip in Lago Maggiore, that his glass pistol produced the most interesting explosions when fired in the presence of a particularly evil smell in the reeds along the edge of the lake. Volta pronounced this smell to be another example of 'inflammable air' – it was methane – and wrote to his colleagues about the possibilities of developing the glass pistol and the gas as a new kind of glass bomb. The eudiometric pistol became very fashionable as interest in hygiene waxed –

Volta's electrophore. The wax and turpentine cake generated static electricity when rubbed with the cat skin; this was transferred to the metal lid with the vertical glass handle. The lid was then, in effect, a portable battery.

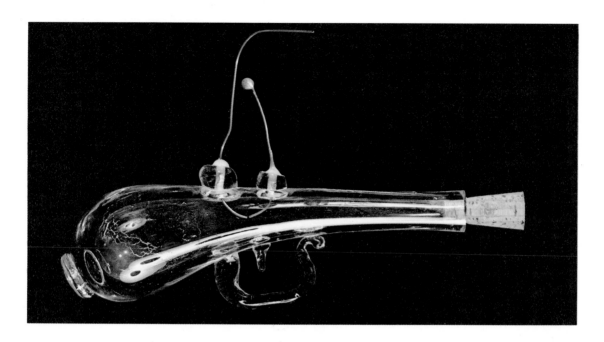

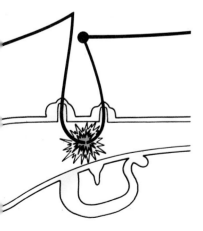

Volta's eudiometric pistol. When the pistol was filled with gas and corked, the experimenter earthed one wire by touching it while making contact between the other wire and the electrophore lid. This produced a spark (see below) where the two wires almost met inside the pistol. The eudiometer was the precursor of the modern automobile sparking plug.

Napoleon even established a eudiometry company in order to produce a bad smell map of Egypt during his campaigns there in the early nineteenth century. The theory prevalent at the time was that malaria was caused by bad air (*mal'aria* is the Italian for bad air), but when it became clear that this theory was untenable interest in eudiometry declined, and with it died Volta's pistol.

It was, however, to play a vital role in the development of one of the modern world's most ubiquitous inventions, thanks initially to events in the Arctic Ocean in the middle of the nineteenth century. By 1800 the American whalers were having to go as far as the North Pacific to catch their dwindling supply of whales. More than 700 vessels were engaged in this enterprise, because the whale was in demand for the oil which was extracted from its blubber as a means of lighting. By 1840 the Greenland whales had almost all gone, hunted to depletion, and the cost of going as far as the Arctic was pushing the price of whale oil so high that there was urgent need for an alternative source of illumination. Coal gas was already in use, but there were vast areas outside the cities in America and Europe where it was not used, and it was for these homes that alternative lighting was being sought.

In 1859 it was found, at Oil Creek, Pennsylvania, by an ex-railroad conductor and bogus colonel called Edwin Drake. He had been ordered there to drill for a substance which had been sent to a group of Bostonian financiers by a salt miner, for whom this black sludge was merely a pollutant. The sludge turned out to be petroleum, and it killed the whaling industry almost at a stroke.

The new lighting-oil manufacturers sold every gallon of oil they pumped out of the ground, with the exception of the volatile, lighter parts of it. These were no good as illuminants, and for a time were thrown away as waste. They were produced in the first stage of distillation of the crude oil, and were considered dangerous because of their low flash point – they would explode at low temperatures. But towards the end of the nineteenth century several developments came together to turn this dangerous waste product into a vital asset. One was a use which had been found for coal gas. In 1860 the Frenchman Joseph Etienne Lenoir had found that a mixture of gas and air would readily explode, and he used this discovery to design a machine not unlike a double version of Newcomen's engine, in which the piston was moved by the explosion. The success of this machine interested a German engineer, Nikolaus Otto, who by 1876 had designed the now famous Otto engine. Supplies of town gas and air were fed into a cylinder and caused to explode. This forced the piston down, and in doing so caused a mechanical linkage to operate a cam. On the return to its original position the turning cam pushed the piston back up the cylinder, expelling the burnt gases. On its next stroke downwards the piston sucked in a new supply of gas and air, and as it came back up compressed the gases. At the point of maximum compression the gases were exploded, and the cycle repeated. It was the classic four-stroke Otto cycle.

This new source of power caught on immediately, and was used to power pumps, sewing machines, printing presses, lathes, looms and saws among other things. Its serious limitation was that it had to remain connected to a gas supply. Without a source of gas it would not operate, and such sources usually only existed in towns. Moreover, the new engine was heavy and extremely noisy. In 1882 one of Otto's engineers, Gottlieb Daimler, and a designer called Wilhelm Maybach left Otto's factory after a series of disagreements, and set up on their own at the village of Bad Cannstadt, outside Stuttgart. It was at this point that the rest of the pieces of the jigsaw of invention were fitted together. Priestley's soda water and his work on gases had roused interest, as has been said, in bad smells and hygiene. Partly as a result of this, attempts had been made to dispel bad smells by good ones; the main technique used was a development of work that had been carried out in 1797 by an Italian hydraulics engineer called Giovanni Battista Venturi, back in the days when interest centred on water and its uses as a source of power. He had discovered that, if water was passed through a tube which was waisted, as the water went through the constricted area its velocity increased, and in order to accommodate this its pressure fell. As the flow emerged into the section of tube after the waist, pressure and flow returned to normal. This discovery was eventually to be used for metering the flow of water in town supplies, but its use in Daimler's context was considerably more influential. The medical profession, interested in getting rid of bad smells, had turned to the scent-makers for their adaptation of Venturi's discovery, the scent spray, and this device was in wide use throughout Europe and America. In the late 1870s a 'scent spray burner' appeared in domestic heating appliances, as a means of using Venturi's principle in heating.

Two views of Rouseville, Pennsylvania, show the rate at which oil towns grew. In 1863 (left) there were just a few buildings and derricks. By 1867 the town had mushroomed (below), with churches, hotels and a downtown business district. Note the higher derricks used as the drillers went deeper for oil.

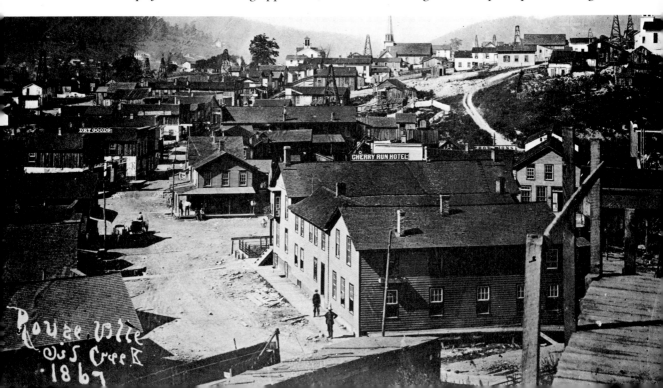

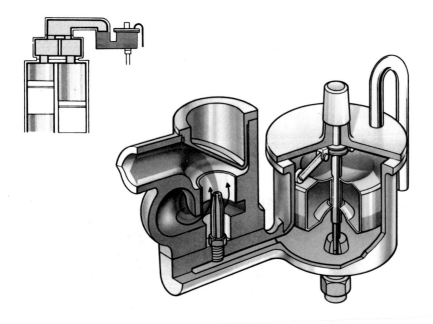

The idea was to use the drop in pressure of an airflow to make it atomize a jet of liquid oil. As the airflow was narrowed, a jet of oil was introduced; the drop in air pressure sucked the oil out of its tube and broke it up into a fine mist of tiny droplets. Whether or not Maybach had heard either of this burner or indeed of Venturi's work, this was the principle he used, together with Otto's cycle of piston movement. The fuel was already available: the volatile fractions of oil which had previously been regarded as too dangerous to use in lighting. Maybach introduced a jet of petrol to the middle of a waisted airflow, the low pressure drew out the petrol and it was atomized into a fine spray ideal for ignition in an engine. His carburettor of 1892 was an instant success, and there was legislation in Germany restricting use of other methods. By 1900 he and Daimler had it fitted to an engine that ignited the fuel with an electric spark, as Volta's pistol had shown was possible. The engine was named after the daughter of Daimler's principal distributor, Mercedes. The new car won every race in the calendar.

Early on in the development of the carburettor Daimler had fitted an engine to a boat, and it was known that he had been trying out the new motorcraft on the German lakes. It therefore came as no surprise to the company when an order was received for the new lightweight engine from a piano-maker living on Lake Tullnerbach, in Austria.

Maybach's spray carburettor. The larger diagram shows how the weight of the fuel in the reservoir (right) forces the fuel up through the vertical nozzle, where it is atomized by the air flow (shown by the arrows). The carburettor (red) is shown top left, located next to the engine cylinders where the explosion of the mixture takes place.

His name was Wilhelm Kress, and in 1901 he ordered an engine that was to weigh 500 lb and develop 40 horsepower. For the previous thirty-five years Kress had been using whatever engines were available to achieve high speeds on water, and he had generated enough interest in the models of the craft he was trying to perfect for him to receive the sum of 5000 crowns from the Emperor Franz Joseph. It was this prize that financed his purchase of the lightweight engine.

As the great day of his first trial with the new engine came, in October 1901, Kress did not know that due to an oversight the Daimler company had delivered an engine almost twice as heavy as his specifications. His twin-hulled craft raced across the lake, engine roaring. On the point of realizing his life's ambition, Kress suddenly saw floating debris in his path. As he desperately tried to lift his machine over it, the craft slewed, capsized, and broke up. By the time Kress was able to repeat the test he had been overtaken by events and denied the privilege of being two years ahead of a couple of American bicycle mechanics called Wright. The machine Kress had so much wanted to take out across Lake Tullnerbach, and which – had the engine been what he thought it was – might have left the water, was an aeroplane.

Wilhelm Maybach (in the white suit) in a Mercedes built by himself. The photograph was taken, in 1903, in the courtyard of the factory at Bad Cannstadt.

763

NOORD

QVIBIRA

LA FLORIDA

MAR NEGRO

MAR DEL·SVR

MAR PACIFICO

7
The Long Chain

The jet aircraft has probably done more than any other modern product of science and technology to bring change to the global community. Whereas the telephone and the computer have broadened the community's mental horizons by bringing its members in contact with cultures different from their own, aircraft have made possible physical contact. In this the aircraft is the direct descendant of the bicycle, which put villages in touch with each other more cheaply and more easily than any previous form of transport, and of the railway and automobile which made contact possible between one country and another. The aircraft has made such contact possible between continents.

It has undoubtedly changed the concept of distance. When modern airline passengers take off they leave the reality of their surroundings, pass a period of time in a travelling capsule, and return to reality at the other end of the flight: the reality of the terrain and the ocean that lies between the point of departure and destination is removed. With the advent of supersonic flight, the concept of time has also changed. Now travellers flying from east to west may arrive at their destination before they have left the point of departure, and in so doing experience the one bodily condition created by the development of the aircraft – jetlag. The increasing use of aircraft has also contributed to the speed with which the world is using up one of its scarcest resources. In one hour's flight hundreds of gallons of fuel are burned, and since the efficiency of an aircraft is measured by the amount of time it spends in the air, the incentive is to get it back into the air with the minimum delay.

The great Chart of the South Sea, drawn in 1622 by the Dutch hydrographer Hessel Gerritsz for the Dutch East India Company. Note that Australia and New Zealand had not yet been discovered. This was one of the principal maps consulted by Abel Tasman, on the voyage in 1643 during which he discovered Tasmania, Tonga, and the Fiji Islands.

The jet has radically altered the rate at which Western technology and culture has spread. When a ten thousand foot runway is constructed in some hitherto remote spot, life in that area will begin to change at once – probably faster than at any time since it was last invaded in war. And because the modern traveller demands as far as is possible the comfort and lifestyle of his home, it becomes increasingly difficult to wake up in a hotel anywhere in the world and know, instantly, where you are. Local customs and the physical shape of the environment are changed to meet the requirements of visitors. The rate at which this is happening has increased with the construction of aircraft capable of carrying hundreds of people halfway around the world. As the standard of living in the West rose steadily after the end of World War II, with a consequent increase in disposable income, the

The original version of the Boeing-747 was 231 feet long with a 195-foot wingspan and had a range of over 6000 miles. In the cargo version the entire nose section opened upwards to permit container loading of 100 tons of freight.

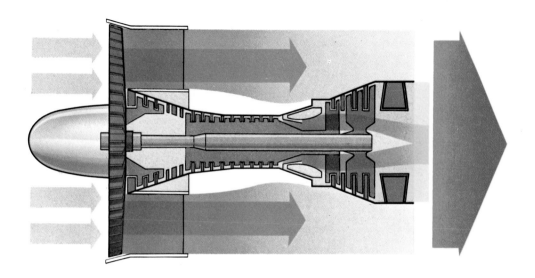

The jet engine. Angled blades mounted on a central spinning shaft suck in air and compress it; the air is then heated in the combustion chamber. Some of this air is used to drive the central shaft while the rest is expelled rearwards through the nozzle to produce thrust.

numbers of people using aircraft for business and pleasure rose too. Because the new jets operate more efficiently than their predecessors, costs have fallen and this has acted as a further stimulus to travel.

It is not surprising that these aircraft were initially designed and built in the United States, where distance has encouraged the use of aircraft since the beginning of the century. The American aeronautics industry is the most sophisticated and extensive in the world, to match the country's position as the greatest trading nation. As such America needs a large, efficient and flexible fleet of aircraft. The widebody jet is by no means the largest airframe that technology can build, but it is the optimum size for the work it has to do. The design embodies no radically new major components; the frame needs regular maintenance of a minor nature only; the replacement rate of its major structural components is measured in decades. In terms of the relationship between cargo carried (passengers and goods), distance covered per hour of flight, minimum time for reloading and refuelling between flights, and number of crew needed, the widebody jet is probably the most efficient form of transport since the seventeenth century. Then, a similarly optimal form of transport was designed and built for the greatest trading nation of the day, for the same commercial reasons. And in terms of our present concern with the effect of jet transport on our own future, with its use of resources, with the effect it will have on the countries which supply those resources, with its contribution to the growth of the economy, the seventeenth-century situation was uncannily similar.

The greatest trading nation at the time was Holland. Throughout the sixteenth century she had been slowly developing a monopoly of Baltic trade. As the European economies had grown, the use of wood had despoiled much of Europe of its great forests, and by the end of the sixteenth century legislation had been enacted to try to slow down the rate at which the woods were disappearing. As their trees went, the French, the Spanish, the Portuguese and the English turned to the nations on the Baltic for alternative sources of supply. Wood was needed above all for the construction of ships, both mercantile and military, and it was this trade which the Dutch had come virtually to monopolize. One other factor is significant in what the Dutch did next. In the last decade of the sixteenth century, as a result of the uprising against their Spanish masters, the southern part of the Netherlands was almost totally destroyed – factories, ports, towns – and refugees from the south streamed north to Amsterdam, bringing with them their skills and their money. The opportunities for them to settle on the land were extremely limited, so most of them entered commerce. Many emigrated to other European cities, to set up trading links with home. As a result of this massive influx, occurring at the same time as the economy rose to a peak because of the Baltic monopoly, the Dutch found themselves the richest country in Europe, with a capital city that was rapidly becoming an entrepot of unparalleled size. Goods flowed into the country from abroad, and then out again as re-exports, with the Dutch taking the middleman's cut. Right in the middle of this early trading wave came the new ship, designed to serve the Dutch commercial machine. It was called a *fluyt*, and the first hull appears to have been laid in 1595 although full-scale production had to wait until the turn of the century. By 1603 there were eighty fluyts at sea, and the number was rising fast.

The reason the Dutch built this particular type of ship when none of their commercial rivals in Europe did so lies in the fact that the Dutch alone did not follow the general trend in ship design, which may be said to trace its origins back to the fall of Constantinople in 1453. Once the city had passed into the control of the Turks, the price of passage for goods and commodities through Constantinople from the east rose considerably. This encouraged the Portuguese to seek their own routes through to the Red Sea, and led to them rounding southern Africa in 1495, reaching and setting up trading posts in India. This precarious foothold in the east stimulated demand for profitable luxury goods like spices, silk, calico, gold and silver, and by the early part of the sixteenth century trade with the East Indies was regular, while in the west the Spaniards were opening up central America and the West Indies for sugar and silver. The great European transoceanic expansion was about to begin.

Then, in the second half of the sixteenth century, that expansion was delayed by war between Spain and the Low Countries – a war begun in the Netherlands, then spreading to France and Spain, and later, as religion entered the argument ranging Protestant against Catholic, involving the entire continent. By the end of the century, when military and economic exhaustion forced a pause, trade began once again to flourish in the period of truce. This time, however, the merchant ships did not sail alone. The oceanic voyages were too valuable to be left to fail at the hands of freebooters, as they so often had when the English had lain in wait for the lumbering Spanish silver ships to emerge into the Atlantic on their way home. Now when the ships went they sailed in convoy with a warship escort. This interest in the fighting ship obsessed every nation but the Dutch.

The sixteenth-century Portuguese carrack, the first three-masted ship in the Mediterranean. Like most ships of the time it functioned both as a merchantman and warship. Note the extremely high stern-castle, needed to accommodate a large crew: these ships carried up to eight hundred men and a hundred guns.

The latest fighting ship at that time was the galleon, first built at the Venice Arsenal – arguably the finest contemporary shipyard in the world, where production-line methods were used and the ship was broken down into component standard parts that were made separately and fitted together in the dock. The new ship was full-rigged, with three or four masts carrying square sails to catch the steady trade winds on the way out and back across the Atlantic. A triangular sail on the bowsprit, above the bows, and another at the stern gave the ship tacking capability in the contrary winds that were found at the beginning and end of the voyage. There were defensive 'castles' rising fore and aft high out of the water. The hull was narrow, and extensively strengthened to support the guns, ranged in one or two tiers on two of the three decks. The defeat of the Spanish Armada had shown that victory lay with the heavily armed ship, and the design of the galleon permitted maximum armament. The bigger ships weighed over 800 tons, and with a ratio of one man per ton they carried an extremely large complement of men, most of them concerned with the guns and the sails. The development of heavier cannon made it possible to sink bigger ships, but at the same time required larger ships to carry the added weight. Some of the galleons carrying heavy armament weighed over 1000 tons.

In every country but Holland, ship design was the responsibility of the Navy, and ships were built that could function both as merchantmen and warships. Thus it was the galleon that brought back the spoils of New Spain, although when the guns and the men to handle them were aboard, there was relatively little space for cargo. For the highly profitable cargoes on the oceanic run, however, the galleons sufficed. Many of them ran a triangular course: down to West Africa with textiles to exchange for slaves; across with the slaves to the West Indies, where the slaves were exchanged for sugar; then home with the

A sixteenth-century view of the Venice Arsenal, the shipyard of Zacharia d'Antonio. The Venetian state was able to maintain a navy of over a thousand galleys due to the production-line methods employed here. Note the ship being launched from the dry dock in the background and, on the left, fitters working on the superstructure of a ship already afloat.

The Dutch fluytschip, first built at the northern Dutch port of Hoorn. In terms of what it was designed to do, this was one of the most efficient ships ever built.

sugar. But the galleons were too big for really profitable trade in European waters, and it was here that Dutch policy paid off.

The Dutch had their warships built by the Navy and their merchant ships designed by traders. These traders designed the fluyt, and in doing so they built a ship that was to be the last word in bulk carriage for two hundred years. The first one was launched at Hoorn, and in almost every way she was unique. The fluyt was unusually long, the length being up to six times the width – twice as long as the standard. The deck area was clear save for a small deckhouse aft, leaving most of the space for cargo. The ship's bottom was almost flat, which permitted a large, almost square hold, easier to fill efficiently. The modified hull design enabled its post structures to be nearly vertical. The sails on the ship were smaller, and the mast shorter than usual, and blocks and pulleys were extensively used to make it easier to handle the sails and to cut down on the size of the crew. As a result of all this, the centre of gravity of the fluyt was lower than normal, and this gave the ship extra stability in rough weather. Many of the fluyt's features were the results of decades of gradual improvements in the boats that sailed the shallow Dutch inland waters, and they created a ship that was rarely over 500 tons in weight. Though technically the Dutch could have built up to 2000 tons, the fluyt's weight was optimal for the cargoes she carried and the distance she travelled. And with a small crew, and the use of pine instead of oak in the upper works, construction and running costs were much cheaper than other ships at the time, which encouraged more frequent sailing times. The fluyt built the Dutch mercantile empire, ploughing mundanely around the European coasts, carrying grain, timber, iron, fish and furs from the Baltic, salt and wine from Spain, Portugal and southern France, woollens from England. Amsterdam became the richest city in Europe. It was not to remain so for long.

Across the Channel, England's colonies had begun to pay off, and by the middle of the seventeenth century were returning extraordinary profits and turning the English into a nation of sailors. Of a population of about five and a half million it is estimated that about fifty thousand were at sea. Shipping employed five times the percentage of population that it does today. The cause of this sudden maritime expansion was initially tobacco and sugar. (It is said, incidentally, that this was when the English developed the taste for sweet tea which they have never lost, because tea and sugar happened to come, from east and west, at the same moment in history.) The colonies were to make England rich. As one merchant put it, 'The purpose of a colony is to take off our product and manufactures, supply us with commodities which may either be wrought up here or exported again, or to prevent fetching things of the same nature from other places for our home consumption and employ our poor and encourage our navigation.' He had put his finger on the reason for England becoming rich, and for the City of London taking on the unique trading position for which it is still known today. That reason was re-exportation of goods. In 1615 the consumption of tobacco in England had been 50,000 lb. At the end of the century the country was importing over 40 million lb, 25 million of which was being sold on the Continent. In that period tobacco prices had fallen four hundred times. During the same time, West Indian sugar imports increased eight times, and calicoes from India quintupled. Over 90 per cent of these latter commodities were re-exported to the Continent. The major problem created by all this long-distance trading was that it needed financial backing. At first this was haphazard, involving only those who had the money to back an entire voyage alone. But there was a limit to what single investors could do, and by the end of the seventeenth century the limit had long been reached. Some new way of generating the necessary money had to be found.

*Ships of the Dutch East India
Company at anchor in the
Chinese port of Canton, 1669.
The Dutch were the first to
follow the Portuguese to the East
in search of silk and spices, and by
the 1620s they had a virtual
monopoly of this trade to Europe.*

It was found in land, which was now registered for the first
time. With registration came regulation in ownership, and on that basis
landowners could confidently borrow against their holdings. These
early mortgages financed England's growth, and they were arranged
in London at places which had been opened to sell the new 'black
liquor' from Turkey. In these coffee houses you could borrow money,
lend it, invest it, or spend it. Some of the coffee houses printed news-
sheets carrying the latest information on imports arriving at the
docks, disasters at sea, city gossip, advertisements from job-hunters
('A young man that is sober, understands Latin, French, writes all
hands and understands merchants' accounts, plays on the violin and
flute, wants an employment'), even dating services ('I now know of
several men and women whose friends would have them matched,
which I'll endeavour to do as from time to time I hear of such whose
circumstances are likely to agree'). Everybody in business, the arts, the
universities, shipping, the aristocracy was to be seen in the new coffee
houses, where you could also buy anything from fish, to girls, to land
in Virginia. The amount of money tied up in transactions soon
demanded some kind of regulation, because it was not uncommon for
a ship to be delayed in transit and for debts to fall due without adequate
cash to cover them. By 1683 the bill of credit solved the problem,
since it could be held against the arrival of the ship and used instead of
cash to pay off creditors. The organization set up to handle the
financing of these bills began life in a tavern called the Nag's Head, in
Cateaten Street, and in 1694 it became known as the Bank of England.

With the number of ships increasing every year, and the length of voyages on the oceanic runs often lasting up to a year – with the risks multiplying the longer the ship was away from port – it was inevitable that something should be done about insurance. There was after all no point in investing in a voyage if the ship went down and all the investors' money was lost. In 1688 a man called Edward Lloyd opened a coffee shop where insurance was bought and sold, and where regular shipping news was read out to the assembled company. By 1700 he was publishing a shipping list which gave the ship's name, the name of the master, the ports of departure and destination, tonnage, number of decks, guns if any, where and when the vessel had been built, the name of the owner and, most important of all, the state of the hull and equipment. The hulls were classified according to their soundness by the letters A E I O and U, and the equipment was described as either G (good), M (middling) or B (bad). The best risk therefore was a ship described as AG, and the worst, UB.

It was this concern for ships' hulls that was to lead, within a hundred years, to an invention that is present in almost every modern home. As the ships sailed more often into tropical waters, their wooden hulls were attacked by a tiny mollusc called *teredo navalis*, which lived in those waters, and which bored into the hulls with devastating

Above: A London coffee house of 1705. Candles, as well as providing illumination, were used as time-keepers during auctions: the bidding ended when the candle burned down to the point where a pin stuck in the side fell out. From this comes the phrase 'to hear a pin drop'.

Right: Ship repairs during a voyage along the African coast in the late fifteenth century. Note the pitch being distilled over the fire on the right. Pitch not only protected the wood from teredo navalis *but also gave extra waterproofing and longer life to the hull.*

results. The only protection against the mollusc was a thick layer of a mixture of tar and pitch smeared over the bottom of the ship. At the beginning of the eighteenth century most of this material came from Scandinavia and the Baltic, from the unit of Sweden and Finland joined under the Swedish crown. Over the previous two hundred years most of Europe had become increasingly dependent on northern timber, as the forests of England, France, Spain and Portugal had become more and more depleted. The timber was used to build ships and to produce tar and pitch. The best kind of wood for making tar and pitch was pine, which was cooked slowly in pits until the tarry substance ran out of the charring wood, to be collected and distilled and then shipped in barrels. This was the trade over which the Dutch had had control, and which had set them on the road to empire. The effect of the Finnish supplies and the route they took to England was twofold: it enriched the Dutch middleman, and it made England dependent on the Finns. In 1700 the Russians, whose northern ports froze over in winter, decided they needed a warm-water port on the Baltic, and moved against Sweden–Finland. The war that followed totally disrupted supplies. Fortunately for the English, there was one other source of supply – which they owned – in the new American colonies; in 1705 Queen Anne's ministers in London passed the Bounty Act, which offered a subsidy of 5 to 10 shillings for every barrel of pitch and tar the colonists of New England and the Carolinas could produce, and government representatives were sent out 'to convince the inhabitants how necessary it is to assist the views of the Mother Country'. By 1725 four-fifths of the tar and pitch used in England came from the American colonies, and when the war in the Baltic ended, American supplies were well enough established for England to do without the Baltic connection, save for small amounts of the highest quality. This was to prove unfortunate, because in 1776 the American colonies revolted, declared independence, and the supply of pitch and tar stopped dead. But the gigantic increase in shipping had already triggered another chain of events that was to provide what looked like an easy solution to the problem. The attempt at the 'easy solution' was to ruin a Scottish nobleman, and lead to one of the great inventions of the Industrial Revolution.

Back in the sixteenth century, when the English shipbuilding boom had begun, laws had been passed to restrict the use of the forests to the Navy shipyards. One of the effects of this law had been to force glass and iron workers to look for alternative sources of fuel for their furnaces. The fuel they found was coal, and it became known that under certain circumstances this coal would produce a tarry liquid. The reason that little had been done about this discovery was that at the time it would have cost more to produce than to import from the

American colonies. The situation was now very different, however, and so Archibald Cochrane, ninth Earl of Dundonald, set out on the path that was to lead to his downfall. He had had an unfortunate start in life, thanks to previous earls backing the wrong king from time to time, gambling excessively, and generally frittering away the family fortunes. When Archibald took the title, in 1778, there was virtually nothing left of the ancestral possessions but the family home, Culross Abbey, near Edinburgh, and a small coal mine nearby. Cochrane decided to try to recoup the family fortunes, and looked for a profitable way to use his coal mine. He knew there was a market for coke in the iron-making industry, so he started by selling coal to the nearby Carron Works. Then he found, through expensive dabbling financed by three marriages, that he could make more from what the coal produced while it was being turned into coke than from the coke itself. Among other things it produced salts which were needed in the process of making alkalis and soap. Another salt was useful in the cloth trade, as an astringent for ridding wool of its natural grease. Cochrane attempted to sell a sticky substance that also came out of the coal as a gum for use in the printing trade. He even went into ammonia production. All of these ventures plunged him deeper and deeper into financial difficulties.

William Murdock's apparatus for producing gas from coal. This is the equipment which was installed at the Phillips and Lee works in Manchester. The container (E) had a capacity of 762 kg.

But the most expensive one of all occurred when he attempted to produce pitch and tar from the coal by a process called destructive distillation. This involved cooking the coal, and condensing the tarry substances out of the vapours that rose from the coal as it cooked. Cochrane covered himself with a carefully worded patent he was granted in 1781: '. . . a method of extracting or making tar, pitch, essential oils, volatile alkali, mineral acids, salts and cinders from pit coal'. He set up a tar works in the grounds of his home, using four kilns capable of taking a total of fourteen tons of coal. He had had to borrow heavily to finance this grand design, and was happily turning out pitch and tar at a great rate, when he heard to his chagrin that the Admiralty had switched to sheathing ships' bottoms with copper. The earl was ruined, and was eventually to die penniless in a Paris slum, in 1831. Ironically in the same city and the same year a relative of his also died, leaving a fortune. Neither knew the other was in Paris.

The other irony of Cochrane's life happened not long before he gave up the tar-making venture. He had accidentally let the pressure in a kiln rise beyond the normal limits, and the kiln exploded. Cochrane gloomily noted that the vapours from the explosion ignited and burned brightly, but did no more than play with the vapour, attaching a gun barrel to a kiln and shooting flames about. He entirely missed the point of what had happened. In 1782, on one of his last journeys to London in another fruitless attempt to interest shipbuilders in his tar, he passed by James Watt's house outside Birmingham, and mentioned the vapours to him. At that time Watt and his partner Matthew Boulton, who were already building steam engines, had hired a young man called William Murdock to act as their agent and manager of their steam-engine interests in Cornwall. (Murdock had apparently got the job because at the interview he had worn a wooden hat which he had turned on a lathe.) In 1791 Murdock obtained a patent for exactly the same work with which Cochrane had failed, 'A composition for preserving ships' bottoms', and in 1792 he was reported to be experimenting with the use of coal to produce vapours that would ignite. His equipment was simple, consisting of an iron container in which the coal was heated, and a pipe leading away to a room which was illuminated by the light from the burning vapours. In 1799 Murdock came to the Boulton and Watt Soho factory in Birmingham to try to interest Watt in his new gas, of which he claimed to be the inventor. Watt advised caution until all doubt was removed regarding the validity of a patent, and then, on the signing of the peace treaty between England and Napoleon at Amiens, Murdock celebrated by installing two gas burners at each end of the Soho factory. The smell they made became known as the 'Soho stink'.

The Boulton and Watt works at Soho, where the two men set up in partnership in 1774. By 1800 the works had turned out some 500 steam engines, of which 52 were in mines and 84 in cotton mills.

By 1805 fifty gas lamps had been made for a firm of cotton spinners, Phillips and Lee, in Manchester. The cotton mills were a ready and willing market. With recent innovations such as the Crompton mule and the Arkwright loom trade was booming, but the number of fires caused by the use of candles and whale-oil lamps was forcing their insurance premiums to unacceptably high levels. Moreover, difficult political relations with the United States interrupted the supply of whale oil, increasing its price. The fifty gas lamps were installed in the Manchester factory in 1806 – and there was no 'Soho stink'. Not only was the new lighting safer, it was also cheaper. It was estimated that instead of the £2000 a year spent on candles, gas illumination over a similar period would cost only £600. The potential market for the new gas was too great to escape the notice of the promoters. Since 1802 a flashy entrepreneur, Frederick Albert Winsor (a German who changed his name from Winzler) had been in on the act, in more senses than one.

Winsor too had been trying to make gas, though in his case from wood, and according to his memoirs the idea of illuminating Britain by gaslight struck him, in a somewhat mixed simile, 'like an electric spark'. He arrived in London in 1802, and by the end of the following year was giving amazing illustrated lectures on gas at the Lyceum Theatre. He did the demonstrating, but because of his atrocious English he hired an Englishman to read his statements. Winsor's style was florid, as can be seen from the innumerable pamphlets he produced in order to promote himself and the gas he claimed to have invented. One of his better efforts reads:

Winsor's burners, whose extravagant designs were produced to catch the eye of potential investors. At one point Winsor promised a return, on a £5 share in the business, of an annual income of £570!

Most mortals on earth with smoke live in strife,
And many a beauty is smothered alive.
Great London itself the emporium of the world
In clouds of black smoke is constantly furl'd.
Smoke begot chimneys, chimneys beget smoke.
Soot, fires and filth, *all prevented by coke.*

His show consisted of an extraordinary variety of burner shapes – chandeliers, cupids, fishtails, heraldic arms, and so on – which flamed into life at the touch of a taper.

However erratic and pompous this ludicrous figure may have been, his apparatus for producing and distributing the gas proves the soundness of his approach to the business. It consisted of an iron pot, closed at the top with clay to make it airtight. A pipe in the centre of the lid covered the cone-shaped condensation vessel, which was divided into compartments by plates perforated with holes to spread the gas and purify it. The gas was then fed to the burners by a central supply pipe. Winsor had seen that the Boulton and Watt idea of supplying gas to houses or factories on an individual basis was not the way to supply towns, which needed large centralized supplies. This would obviously demand immense amounts of capital, which Winsor set himself the task of obtaining. In 1806 he was canvassing for funds

Gaslight in the streets of London created the new occupation of lamplighter. A contemporary poem written in honour of the task ran: 'He still will strive on every coming night, To send you home with artificial light'.

for a National Light and Heat Company to provide 'our streets and houses with light and heat . . . as they are now supplied with water'. In 1807 he displayed lights along the top of the wall between Carlton House and The Mall. The money began to pour in, and in 1812 the company was renamed the Gas Light and Coke Company. In 1814 came Winsor's finest hour. He built a gas pagoda in St James's Park to celebrate the visit of the Allied Sovereigns, *en fête* after Napoleon's banishment to Elba, and the night before the great display had the opportunity to show it off to the Prince Regent. This turned out to have been a stroke of fortune, because the next day one of the courtiers insisted on lighting fireworks on the structure before the gas exhibition began. There was a deafening explosion, and the entire pagoda went up in flames of a less controlled nature than Winsor had intended to display. He had made his point, however, and with widespread support he lit the Parish of St Margaret's Westminster in the same year. Within six years the major British towns were illuminated by the new gaslight. By 1823 there were 300 miles of mains piping, and by 1850 this had increased to 2000 miles. The gas created a sensation wherever it went. Sir Walter Scott was not impressed. He wrote to a friend: 'A madman is proposing to light London with – what do you think? Why, with smoke!' But in general reaction was favourable, as witnessed by a letter to Lady Mary Bennet in 1821: 'What folly to have a diamond necklace or a Correggio, and not light your house with gas! . . . Dear Lady, spend your fortune on gas-apparatus. Better to eat dry bread by the splendour of gas than dine on wild beef with wax candles.'

The social effects of the gas were widespread. It made the streets instantly safer at night. 'Before', it was said, 'the light afforded by the streetlamps hardly enabled the passenger to distinguish a watchman from a thief, or the pavement from the gutter. The case is now different, for the gas lamps afford a light little inferior to daylight and the streets are consequently divested of many terrors and disagreeables.' The gaslight made longer factory hours possible, and in consequence production levels rose. The evenings could now be fully used, and the spread of literacy that followed the setting up of workers' institutes increased the sales of books. The new light also gave birth to evening classes, introduced to teach those who had not had the benefit of education beyond the age of fourteen. Courses were held in a wide variety of subjects from philosophy to mechanical drawing, but many concentrated on meeting the growing need for the basic abilities of reading, writing, arithmetic and grammar. The Mechanics Institutes spread all over the country, providing lectures on science and the arts. The great London clubs grew up as evening became, for the first time in history, a time of day when community activity could take place.

Cremorne Gardens, London, 1864. By this date gaslight had been installed all over the city for more than twenty years. The genteel could now promenade the town squares and parks at dusk without fear of being molested by footpads lurking in the darkness.

Not all the new evening activities made possible by the brightness of gaslight were educational, as this painting of the regular illuminated rat-catching sessions at the Blue Anchor Tavern in Finsbury shows.

The only part of the capital which was not to benefit from all this enlightenment was the river Thames, repository of all the waste products of the coal distillation process, tons of which were dumped into the waters. Similar dumping was going on wherever the new gas-producing machinery was in action. Then in 1819 a Scotsman who ran a cloth-dyeing business in Glasgow arrived at the gas works looking for cheap naphtha, which had been identified in the coal tar waste. He believed that the naphtha could be used as a cleaner for his dyeing machines, and since it was being thrown away he hoped to get it cheaply. At some time in 1820 when he was using the naphtha on various materials, he accidentally found out that it would dissolve rubber. Since it had first been brought back from America by the Spanish rubber had been in limited use, and took its name from the principal one, that of erasing draughtsmen's errors. It was also used to some extent in surgical catheters, and as cores for balls. The new discovery that rubber could be dissolved enabled it to be used in the clothing industry, when the dyer in question spread the dissolved rubber between two sheets of cotton and called the product after himself: macintosh. In 1823 the waterproofing patent was issued to Macintosh, and within a year he was supplying the government and had set himself up in partnership with a Manchester cotton mill.

The dashing all-weather hunter, defying the rain in his hat and cloak of 'paramatta', a rubberized woollen cloth. The new waterproof clothes encouraged people to take part in more outdoor activities.

Unknown to him at the time, an ex-coachbuilder called Thomas Hancock was also experimenting with rubber. He may have been looking for a way to weatherproof his coaches, but whatever the reason in 1819 he was in business in London, selling rubber. He bought the material in 'bottles', the shape in which it arrived from South America, because the latex from a tapped tree was allowed to flow into a glass bottle, and when the contents had hardened the glass was broken. This rubber was cut into strips, which when warmed up would stick together. Hancock's 1820 patent for this method was used to make gloves, braces, stockings, garters, shoes and soles. These articles caught on immediately they appeared on the market, which left Hancock with a problem: he did not have enough rubber, although after he had cut each bottle into strips, he was left with odd-shaped ends he could not use. Eventually he developed a masticator, a box with a roller inside it on which teeth were fixed. These teeth shredded the rubber, so that it could then be pressed and once more cut into strips.

Meanwhile Macintosh's cloth was proving a great success on the Franklin expedition to the Arctic in 1824. Rubber began to appear everywhere, in hoses, printing rollers, beer engines, air mattresses, blankets, life preservers, hats, boat covers and bicycle tyres. In 1825 Hancock and Macintosh got together and began producing the now familiar macintosh raincoat. From then on the growth of the rubber industry was phenomenal. From a total of 23 tons imported in 1830 the amount coming to Britain rose to over 5000 tons in the 1850s. And still Hancock and Macintosh could not get sufficient supplies to meet demand. In 1853 Hancock wrote to the authorities at Kew Botanical Gardens in London to try to get them either to grow samples from seedlings he sent them, or to obtain some from South America, with a view to setting up rubber plantations in the Far East. The development of rubber might have taken a completely different course had those at Kew not been involved at the time in problems relating to another South American plant which also had some significance for events in the Far East.

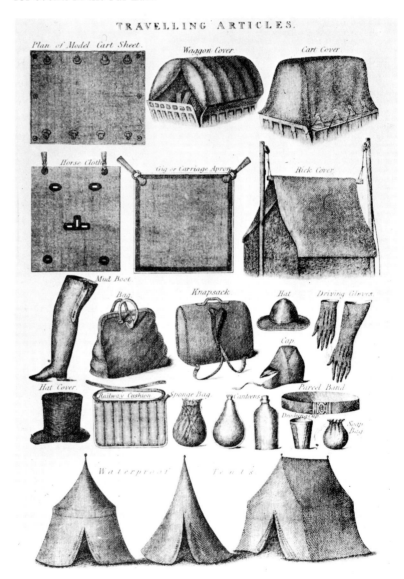

A few of the many articles that were made of rubberized cloth. Although a selection like this may have been of limited use to the individual, it is not difficult to see why it appealed to the Army, required as it was to operate in any weather.

The plant in question was the cinchona, which had first been discovered by Europeans when Jesuit priests had come across it in Peru in 1630. The bark of the cinchona contained quinine, which was used in the treatment of the malaria ravaging the armies and administrators of British India in the middle of the nineteenth century. It was for this reason that the botanists at Kew were desperately trying to obtain cinchona seeds, so that cultivated seedlings could be shipped to India in order to grow cinchona where it was most needed. In 1859 two expeditions were sent to Lima under the botanist Charles Markham. The first got lost; the second managed to return with seeds, but the plants they produced gave only a very weak form of quinine. Meanwhile Dr Forbes Royle, the government botanist in India, wrote home: '... it is a drug which is indispensable to the treatment of Indian fevers. I have no hesitation in saying that, after the Chinese teas, no more important plant could be introduced into India.' Eventually the tiny cinchona plants arrived in Madras, where they failed disastrously, leaving the Indian Office and its vast number of civil servants and military personnel dependent on imports at extremely high cost from Java, where the Dutch had successfully transplanted the same kind of seedlings. The administration in India deluged London with letters demanding that something be done. The service was being decimated by malaria, they argued, and without quinine India could not be governed at all. The life-expectancy of a British soldier, for example, was half what it was in England.

While all this was going on the rubber-makers continued to buy up coal tar, extract the naphtha from it, and dispose of the rest. Gradually scientific interest began to focus on the tar. By the beginning of the 1850s several chemical compounds had been identified in the sludge, one of which was not unlike the chemical structure of quinine. At the Royal College of Chemistry in London a German professor aroused the interest of one of his young assistants in the possibility of altering this compound in such a way as to make quinine from it. The young man was William Perkin, and during the Easter vacation of 1856 he made an accidental discovery. In his own words: 'I was endeavouring to convert an artificial base into the natural alkaloid quinine, but my experiment, instead of yielding the colourless quinine, gave a reddish powder. With a desire to understand this peculiar result, a different base of more simple construction was selected, viz., aniline, and in this case I obtained a perfectly black product.' The aniline to which Perkin referred was the generic name given to all substances derived from coal-tar naphtha. The choice of this material for a second experiment was to make Perkin a millionaire, because when he washed and treated the 'perfectly black product', it became mauve – the first artificial aniline dye. After having undergone hesitant technical trials,

The colourful world created by the dye chemists. An early BASF advertisement for their Indanthren range proclaims the ability of the new dyes to reproduce the colours of nature.

A replica of the Empress's green dress, made by the Hoechst Company in 1977. Its effect on the public when she first wore the dress made green silk instantly fashionable, and the Lyons silk weavers had a monopoly on the colour.

the dye went on sale in 1857 from a small factory outside London which cost Perkin's father every penny he had (Perkin himself had no funds, being at the time only nineteen). Suddenly it was a colourful world. Even Queen Victoria wore the new mauve. Almost immediately the oxidation of aniline was investigated and elaborated by other chemists, and a flood of new aniline colours appeared: first crimson fuscine, and later blues, violets and greens. At the great 1862 exhibition in London, the new colours delighted the crowds.

Meanwhile the raw material for all this – coal tar – was still pouring out, virtually free, from every gasworks in the country, and the race was on to find what else it would produce. In 1863 a German chemist, Eugen Lucius, one of the founders of the giant Hoechst firm, discovered a new green. Its début was spectacular. A Lyons silk-dyeing firm called Renard and Villet bought all the green Hoechst could make, and used some of it on a silk which was then made into a dress for the Empress Eugénie. The night she wore it to the Paris Opéra for the first time it took the fashion world by storm. It was the first green which did not go blue in gaslight! In the early 1870s the German chemists working in Britain, including among them the same German professor who had originally aroused Perkin's interest – A. W. von Hofmann – all returned to Germany, either recalled or having received offers they found it impossible to refuse. At the same time Perkin retired at the age of thirty-six to devote himself to 'pure science', and the British dyeing industry went into a decline.

One of the reasons Britain let slip its early lead in the industry was that the country was rich. It had a stronger industrial base than the rest of Europe, and benefited from cheap imports from the colonies which also acted as captive markets for British manufacturing. It was too easy for money to be made in areas of activity that were already tried and known to be sound. No banker would risk money on a new venture such as dyes when he did not need to. Besides, the industrial scientist and engineer enjoyed a relatively low social position and perhaps because of this, like Perkin, they felt some embarrassment at making money. Like Perkin they dreamed of the day they could afford to return from industry to 'pure science'.

The German attitude was entirely different. The influence of the great German philosophers of the eighteenth century, among them Kant and Wolff, and the pedagogic teachings of the Swiss Pestalozzi helped to alter radically the structure of German education. These men believed that the body of man's knowledge was incomplete, and that the work of higher education was not to inculcate the repetition of known fact but to train minds in the business of investigation. In the early nineteenth century the Germans founded technical high schools, and by the 1880s the standardization of university and technical school training mixed the social classes in a way that would never have been acceptable to the English. Thus the new German dyestuffs industry had only to go to the schools to find the talent on which to build. The firms had close personal ties with the academic staff – encouraged, in the case of BASF, by Heinrich Caro, a chemist who had come back from England to join the research team and whose association with the universities helped towards the success of BASF. His contact with the academic laboratories kept a stream of new colour processes coming in to the BASF factories, and this system was adopted by the other great German companies such as Hoechst, Bayer and Agfa. German expertise with colour was also to lead to discoveries in apparently unrelated fields, such as that of medicine: the investigation of the chemistry of colour led to systematic thinking about the structure and effects of chemicals, and this led directly to drugs like aspirin and to techniques for staining tissue for diagnosis. It was this use of tissue staining to identify potential sufferers from syphilis that led to the disease being treated successfully with the stain chemical itself. The new drug was called Salvarsan.

By this time coal tar had been exhaustively analysed for its constituents, most of which were then in use. About the only thing left unexamined was what had once been called the 'Soho stink'. This too was about to come into its own, thanks to the sudden and electrifying change by which Germany, from being an almost totally agricultural country in 1870, became the powerful industrial giant she was forty years later. During that time the German population rose from 41 million to 66 million, while that of France remained unchanged and that of Britain rose from 26 million to 40 million. The massive leap in Germany was due in part to the rush to the cities that accompanied the growth of the German iron and steel industries. One result of this growth of population was that it caused a food crisis. The trouble stemmed from the aristocratic Junkers of East Prussia, Germany's breadbasket. The Junkers owned much of the land – half of all arable land was in entailed estates – and they grew rye on it. The soil itself was too light for wheat, and besides, rye was fertilized with potash, of which Germany had vast quantities. By 1870 American grain prices were becoming highly competitive, thanks mainly to the invention of the McCormack reaper which could harvest six times faster than a man. This low-price American grain caused a slump in agricultural profits, and the Junkers reacted by demanding – and getting – subsidies. The irony was that they were being subsidized for exporting their rye, and at the same time getting tariff barriers erected against imported grain. So although Germany could have eaten black rye bread, all the German rye was leaving the country, and not enough wheat for white bread could be grown in the absence of fertilizer. At the time, most of the land in Europe was being fertilized with sodium nitrate from the west coast of Chile, where there were large natural deposits. These came in the form of a kind of ore, which was crushed and boiled to produce crystals, which were then pounded to a dust which was spread on the fields. By 1900 the deposits looked close to depletion, and unless alternative supplies were found Europe was in real danger of widespread starvation.

The answer to the problem came from the 'Soho stink'. It was known that this smell was produced by ammonia, which was present in the coal tar. In the middle of the nineteenth century an English agricultural chemist called George Fownes had suggested trying to turn the ammonia into a salt for use as a fertilizer, and this process was already technically possible. The principal difficulty lay in producing the salt in sufficient quantities. Then, in 1909, a German chemist called Fritz Haber, a product of the technical school system, gave BASF a demonstration of a method that would do the trick. Because he had done much of the preparatory work with a BASF employee called Carl Bosch, the process became known as the Haber Bosch process.

The technique relied heavily on extremely strong pressure vessels, which the recently developed German steel industry was able to provide, and on the supply of liquid hydrogen and liquid nitrogen, provided by the new refrigeration techniques developed by Carl von Linde as part of his work in cooling down Munich beer cellars. Haber put the hydrogen into the pressure vessel together with nitrogen in the ratio of three to one, with a pressure of 200 times normal air pressure and a temperature of 600°C. The gases were then passed through a wire mesh made of osmium, which acted as a catalyst to turn the gas mixture into ammonia. This was an entirely new way of obtaining ammonia. The gas was then treated by a method already in use. It was passed through another hot oven, and mixed with air. As it passed through a platinum wire mesh, the mixture combined to produce nitric oxide. When nitric oxide is mixed with water the result is nitric acid, and when soda is added to the acid the result is sodium nitrate – which is the same fertilizer as was originally being imported from the Chilean deposits. Haber's process for making ammonia in large quantities should have solved Germany's fertilizer problem at a stroke. That it did not do so was due to experiments conducted in Paris by a Frenchman called Henri Moissan.

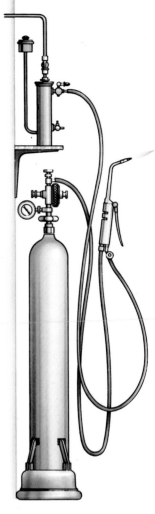

Above: One of the few long-term successes to come out of the acetylene disaster: oxyacetylene welding.

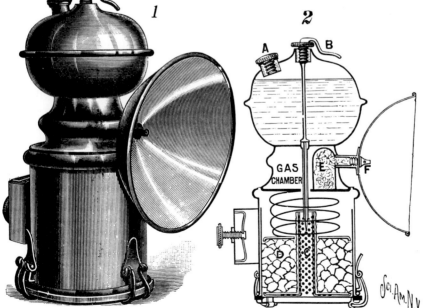

Left: A common use of acetylene in 1900 – in a bicycle lamp. Water from the reservoir dripped on to the calcium carbide beneath, and the gas given off was lit at the nozzle in the centre of the reflector.

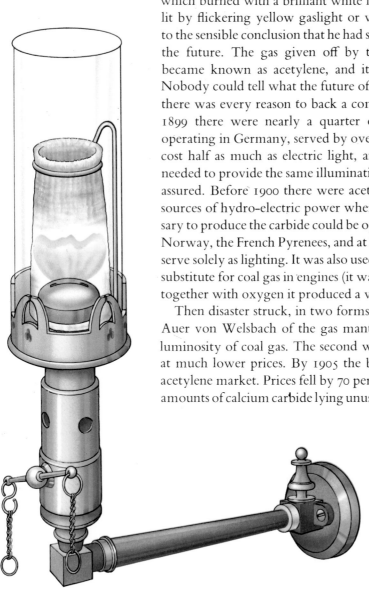

The Welsbach gas mantle, produced in 1893. The mantle was made of silk or cotton impregnated with 99 per cent thoria and 1 per cent ceria. This mixture became incandescent when the gas flame played on it, and greatly increased the luminosity of the gaslight.

Moissan started out by trying to make artificial diamonds, using the new electric arc furnace in which very high temperatures could be produced. It was known that diamonds could be turned into basic carbon in high temperature conditions, and Moissan wanted to see if he could reverse the reaction process and get cheap diamonds. He failed. In 1895, after a number of tests which involved putting almost every chemical he could think of into the furnace, he tried a mixture of lime and carbon at a temperature of 2000°C. The result was a compound called calcium carbide, which gave fairly uninteresting results until he brought it into contact with water, when it gave off a gas which burned with a brilliant white light. Moissan, living in a world lit by flickering yellow gaslight or very expensive electricity, came to the sensible conclusion that he had stumbled across the illuminant of the future. The gas given off by the calcium carbide and water became known as acetylene, and it attracted a lot of investment. Nobody could tell what the future of gas or electricity would be, and there was every reason to back a competitive source of lighting. By 1899 there were nearly a quarter of a million acetylene gas jets operating in Germany, served by over 8000 acetylene plants. The gas cost half as much as electric light, and took up a quarter the space needed to provide the same illumination by coal gas; its future looked assured. Before 1900 there were acetylene plants near several major sources of hydro-electric power where the electric arc furnace necessary to produce the carbide could be operated cheaply: in Switzerland, Norway, the French Pyrenees, and at Niagara Falls. Acetylene did not serve solely as lighting. It was also used to produce lampblack, and as a substitute for coal gas in engines (it was four times more efficient), and together with oxygen it produced a very hot flame ideal for welding.

Then disaster struck, in two forms. The first was the invention by Auer von Welsbach of the gas mantle, which greatly increased the luminosity of coal gas. The second was the production of electricity at much lower prices. By 1905 the bottom had dropped out of the acetylene market. Prices fell by 70 per cent, and soon there were great amounts of calcium carbide lying unused all over Europe and America.

At this point the German dye-makers became involved again, in the persons of Caro from BASF and his colleague Adolf Frank, who were looking for a way to use cyanide as a colour base. They found something else. At one point they ground up calcium carbide, placed it in a furnace at a temperature of 1000°C, and passed nitrogen gas over it. The nitrogen combined with the carbide, and produced a cake-like mixture, calcium cyanamide. This was rich in nitrogen, and it occurred to Caro and Frank that they had stumbled across the answer to Germany's fertilizer problem, because in this form the nitrogen could simply be ploughed into the soil. What is more, the mixture was far cheaper than that produced by the Haber Bosch process. In the face of cyanamide as a fertilizer, the Haber Bosch system failed. It was however to have success in a very different field.

By 1908 German businessmen were concerned at the lack of foreign markets available to their growing industries, thanks to the hegemony of the French and the English in overseas trade. Germany, it was argued, was now an industrial world power, and if she were to take her rightful place in the world she had to have colonies. Only Britain stood in the way. What was needed was a bigger German Navy, to counteract the sea power of the British. Germany began to build more warships, and, in retaliation, so did Britain. This arms race was to continue until the outbreak of the 1914 war, and at that point Germany found herself in trouble of a different kind.

The BASF indigo laboratory in 1900. It was the scientific and industrial expertise generated in organizations such as this that was to turn Germany into an industrial giant with imperial pretensions. By an extraordinary irony, British troops in the First World War wore uniforms coloured with German khaki dye.

The Illustrated London News *of 1889 shows the digging out of caliche at the nitrate grounds in the Chilean province of Tarapaca. These were the deposits blockaded by the British at the start of the First World War.*

After only a few weeks of hostilities, the Germans realized that at most they had enough explosive for a year. This was because the main source of explosive material was the same Chilean deposits that had supplied them with sodium–nitrate fertilizer. These supplies were now blockaded by the British, which is why the first major naval confrontation between the two nations, known as the Coronel, took place off the coast of Chile on 1 November 1914. Without access to these supplies Germany would have to give up the fight by early in 1916. Then somebody remembered the Haber Bosch process, which produced sodium nitrate. Only one more stage was required – to add sulphuric acid to the sodium nitrate. This produced nitric acid which, if poured on to cotton, makes gun cotton, which is explosive. Thanks to its original need for fertilizer, Germany was able to fight on for four years.

One other major development came out of the acetylene débâcle. In 1912, at the German firm of Greisheim Electron in Stuttgart, a chemist called Fritz Klatte was working on acetylene, trying to find a material which would dope aircraft wings with a protective coating impervious to the climate. One of the mixtures he tried was of acetylene with hydrogen chloride and mercury. When this mixture was allowed to stand in sunlight, a milky sludge formed and then solidified. Klatte made a note of the ingredients, and in thorough German fashion filed a patent on it. It was, however, 'not useful' at the time, and further work on it was dropped. Eventually the company let the patent lapse in 1925.

The material Klatte had thrown away was vinyl chloride, and in the 1930s, when it was too late for Klatte to do anything about it, interest in the substance revived. The result of this work was to produce a variant of the chemical called polyvinyl chloride, or, to give it the name by which it is now known throughout the world, PVC – the forerunner of the plastics without which the modern industrialized nations could no longer function.

How the long-chain polymer is formed. Each type of atom seeks to bond with a certain number of other atoms: hydrogen (green) bonds with one other atom, oxygen (brown) with two, nitrogen (blue) with three, carbon (red) with four.

hydrogen – 1

oxygen – 2

nitrogen – 3

carbon – 4

Two molecular groups in which each atom has the desired number of bonded atoms attached.

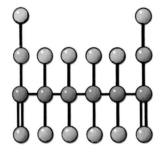
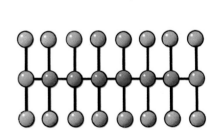

Bringing the two groups together and heating them causes the two hydrogen and one oxygen atom to combine and escape as steam. This leaves atoms at contiguous ends of each molecule lacking the requisite number of atoms with which to bond.

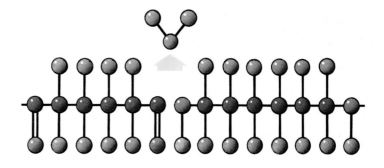

The deprived atoms find their missing bond-mates. This process is repeated any number of times: the new, long-chain molecule produced is nylon.

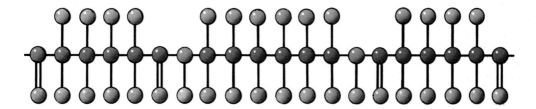

8

Eat, Drink and Be Merry

An early fifteenth-century Burgundian feast. This illustration from the famous Très Riches Heures du duc de Berry *shows the duke and his finely dressed courtiers at dinner. Despite the richness of the setting (note the elaborate gold salt-cellar, right), dogs were allowed on the table.*

The world we live in today is plastic, in two senses of the word. The first is the name of the material with which we are all so familiar. Look for a moment away from this book and at your environment, and see how much of it is plastic. Today, plastic takes the place of rubber in cables and wiring, of wood in furniture and building construction, wool in clothing and upholstery, stone in flooring, oil in paint, metal in automobiles and aircraft, glass in containers, paper in packaging, leather in shoes. Plastic enables us to possess objects which we could not afford if they were made from natural raw materials, and that democracy of possession is recent enough in our society for us still to shape and colour plastic to resemble the materials it replaces. Man-made fibres are treated to look like wool, plastic is grained to look like leather or painted with knots to look like wood. Because we are not yet at ease with the new material we live in a world which although totally reliant on man-made substitute material is made to look as if that material did not exist, so effectively is it camouflaged. We are honest only with innovations that have come since the discovery of plastics. Plastic calculators look plastic, as do fertilizer bags and credit cards.

Plastics have also made our world plastic in the other sense of the word: capable of changing shape with ease. Synthetic materials can be extruded, spun, moulded, injected and pressed, and because of the ease with which they can be handled, manufacturers can and do change the shape of their products frequently and cheaply. Consumers no longer buy simply the object that they wish to possess, but the shape of that object that pleases them. As a result, such alterations in the external shell of goods to suit public taste are used to stimulate a desire for novelty in design. There are fashions where none existed before, and as the plastics era lengthens it becomes easy to see how those fashions have changed. You can, for example, tell at a glance the various changes in style of radios over the past fifty years. The major advantage of this ability to design new models of the same basic object is that it maintains high turnover rates in the market. Consumers will buy a new model simply because it is there. This built-in obsolescence of style does much to maintain high employment levels, since it prevents saturation of the market and a drop in demand. It is an irony that the system by which this world of high turnover, stimulated demand and craze for novelty maintains its financial equilibrium should be embodied in an object which is itself plastic: the credit card.

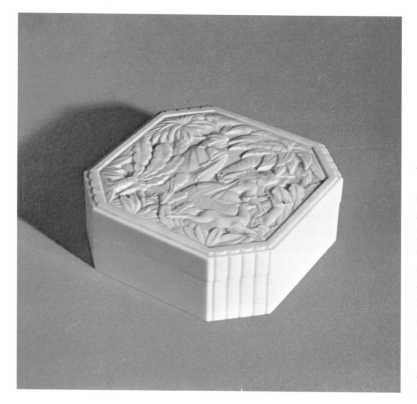

This delicately carved ivory box is no such thing. It is injection-moulded plastic, but you would need to examine it carefully to know that it was made of synthetic material. In spite of the undisputed value in the modern world of such substances, it is only comparatively recently that the word 'plastic' has begun to lose its pejorative meaning of 'cheap and tasteless'.

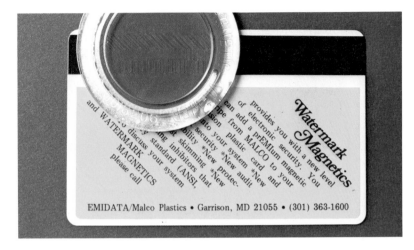

How the code is positioned on a magnetized tape strip, as revealed by a 'reader' – a circular transparent container filled with iron oxide solution. The iron particles are attracted by the areas of the tape which have been magnetized to form a pattern unique to this strip.

The modern world moves too fast for cash, save in personal transactions. Electronic fund transfer makes it possible for money to move as fast as goods, since it no longer actually *moves*. But the credit system has created a new cardinal virtue: creditworthiness. Coded on to the magnetic strip at the back of a credit card or an ATM card may be a totally secure personal key to financial ease of operation, but at what cost? The amount of information which has to be divulged in order to obtain a good credit rating would, fifty years ago, have been regarded as an unacceptable invasion of privacy. Unfortunately it appears to be inevitable. If you wish to buy gas and the pumps accept only credit cards, you either get a card or walk. The question is not, however, simply one of finance. As the amount of data on the individual grows – in banks, police records, schools, places of employment, insurance firms – the security of that information becomes of paramount importance, because access to it on a wide scale would confer immense power. In the meantime what will happen when it is no longer possible to live without a credit card? How can someone opt out of the credit system in a world where cash is not accepted? What social effects would be felt in a cashless society, when the physical presence of cash as a check on falling into debt is no longer there? What happens when every individual lives in deficit?

The first time this possibility presented itself was in the fifteenth century, and the effect was felt throughout Europe. It happened in the Duchy of Burgundy, in France. This territory, which stretched five hundred miles from Holland, in the north, to Switzerland, was ruled by four dukes in succession between the middle of the fourteenth century and 1477. With rare exceptions, the dukes spent much of their lives in debt, thanks to the availability of Italian credit. The bankers from whom they borrowed were the Medici, the most important in Europe at the time.

The Tuscans had been pre-eminent in finance since the thirteenth century, when a Pisan called Leonardo Fibonacchi brought the arabic numerical system back from North Africa, and double-entry book keeping appeared. This made it easier to run complex accounts in an orderly manner, instead of in the jumbled, narrative way it had been done up to then with each transaction written as a separate credit-and-debit account. In the fourteenth century the Italians had also developed the bill of exchange, as a means of handling the increasing international trade of the time without requiring merchants to carry large amounts of cash on unsafe roads. The bill took the form of a credit note for a certain amount, payable at a rate agreed before the merchant left his home, in a specified foreign location for a certain sum of foreign currency. This system needed banks to run it, and these evolved gradually from the counters at international fairs such as those of the Champagne area, where the exchange of money or of credit note for money took place. Monopoly of these techniques had by the fifteenth century given the Italians control of the European money market. By 1470, in Florence alone, there were thirty-two separate banking firms. Three of these operated internationally: the Bardi, the Peruzzi and the Medici.

The Medici established branches in Rome, Naples, Geneva, Venice, Pisa, Avignon, Lyons, Milan and Bruges. The London branch was forced to close in the early fifteenth century because clients, including the king, had withheld payment of debts. In June 1467, when the

Fifteenth-century Tuscan bankers at work, in Prato, near Florence. Large-scale banking began in Florence as a result of the city's extensive trade in wool, although the oldest bank in Europe is the Monte dei Paschi of Siena.

fourth Duke of Burgundy, Charles, came to power, the bank-manager in Bruges was an egocentric, ambitious, would-be international diplomatist and friend of the famous called Tommaso Portinari. Portinari was by then already in trouble. He had fancied himself chief adviser to Charles's father, the third duke, and when the duke decided to build up a navy, Portinari had agreed to finance the construction of two galleys with the bank's money. Charles's father then changed his mind, leaving Portinari with two ships and no payment. All Portinari could do was charter them, at a loss to the bank – much to the annoyance of the bank's president, Cosimo de' Medici. In 1469, when the new president was no less a person than Lorenzo the Magnificent, Portinari signed a new contract of service with Florence. These bank-manager contracts required the manager to put up a certain percentage of the branch funds, while the central bank provided the rest, usually the greater share. In Lorenzo's contractual letter to Portinari one clause said 'deal as little as possible with the Dukes of Burgundy . . . especially in granting credit', and another told Portinari to get rid of the galleys and indulge in no more shipping ventures. It is typical of Portinari's reaction to this very powerful warning that he called the new bank 'Lorenzo de' Medici, Tommaso Portinari, and Co.' The maximum amount the contract permitted to be loaned to Charles was £6000, and Portinari exceeded this maximum. Part of the reason he was allowed to do so lay in the fact that Lorenzo had recently become the leading member in a cartel controlling the new alum deposits discovered on papal lands near Tolfa, in Italy. Alum was a commodity vital to the clothing industry, being used as an astringent to rid wool of its natural grease before it was woven. Lorenzo evidently hoped that his generosity in allowing Charles extra credit would be reciprocated by Charles ceasing to import alum from the Turks. (He needed the alum because part of his duchy included the rich textile towns of Flanders.) Charles, however, took the extra credit and continued to deal with the Turks.

Whatever the reasons for lending money to the House of Burgundy, the practice had been going on since the time of the first duke, Philip, who borrowed the money to buy his first major northern territory, the Duchy of Brabant, and who as a consequence spent the rest of his reign in debt. His son owed a quarter of all his income, and had to pawn his jewels. Charles's father, the third duke, was a little more thrifty. All four dukes, however, raised the credit they obtained by leasing the right to collect rents from their property or revenues from their land, for a specified time, in return for cash in hand or immediate credit facilities. In this it may be said that the Burgundian dukes invented a new way of financing state expenditure. Much of their money went on prestige displays. The Burgundian court was

the most extravagant in Europe, and under Charles the extravagance went to unparalleled heights. His court was the leader of fashion, with its heavy velvets, sumptuous silks, furs covered in jewels, cloth-of-gold, damasks and heavy brocades. The leader in this spending spree was Charles himself, who went to the lengths of having a steel and velvet 'war hat' made, hung all over with diamonds, for wear on the battlefield. Charles saw himself as a second Julius Caesar, the saviour of Europe. He would rather spend a day in armour than write a letter. He would devise elaborate rules for conduct on the field, and then lose the battle. His continuous military fiascos reveal the ambition that drove him to lunacy in his desire to extend his territories at any cost. He saw himself as the natural heir to the imperial throne, and travelled with a crown in his saddlebags so that it would be at hand when the Emperor decided to name Charles his successor. The only time Charles managed to pin the Emperor down was in 1473, at Trier on the Rhine, when he was crowned Duke of Guelders. The Emperor's view of this upstart is amusingly reflected in the words of one of his councillors who was at Trier.

> So the Emperor rose at daybreak on 25th November and hurriedly took ship. Peter von Hagenbach [one of Charles's men] followed after his Grace in a rowing boat and told the Emperor that the Duke was distressed that he got up so early. He had not expected this, and he asked him to go slowly, so that the Duke could come and take friendly leave of him, and talk further about all sorts of things. The Emperor agreed, if he was not too long. So the ships drifted without oars for half an hour. When the Duke did not come, Peter said he would hurry to the Duke so that he might come soon, but as soon as he had rowed away and was out of sight, the Emperor had his oars put out and rowed off.

Charles's ambition and his access to credit were to be his downfall. Thanks to the ability of Portinari to provide cash on the battlefield Charles knew that he could always pay his men, and since this ability was crucial at the time because so many of the combatants were mercenaries who would leave the field if the cash was not forthcoming, Charles rushed into engagements recklessly, and lost them all. In 1476 he decided to strengthen his lines of communication with Italy, the source of many of his mercenaries, by annexing territory to the south of Burgundy. He moved into Savoy with his mercenaries, who included English, Italians and Germans as well as his own Burgundians. The army had with them 100 cannons of various sizes, but the main striking force was the body of heavily armoured Burgundian knights. Behind the army came wagons carrying almost all of Charles's personal jewellery and an immense amount of gold and

Charles the Bold presiding over a meeting of the chivalrous society he founded, the Order of the Golden Fleece. Its members were drawn from the counties and duchies over which Charles ruled: Burgundy, Brabant, Limburg, Luxembourg, Flanders, Artois, Harnan, Holland, Zeeland, Namur and Malines.

silver plate, books, reliquaries, candlesticks, cups, coins – and his throne. The only opposition to Charles's plans for Savoy came from the confederation of towns in what is now Switzerland, which saw Charles as a threat to their independence. So as Charles moved south his first objective was the relief of certain towns along the edge of Lake Neuchâtel, which he had previously garrisoned and which were now under Swiss attack.

On the snowy morning of 2 March 1476 Charles was moving his army along the lake towards Vaumarcus, in three columns stretching back for five miles. Suddenly out of the forest on the hills above him came a detachment of Swiss, blowing their horns and uttering blood-curdling yells. Charles's cavalry attacked. The result was to shake feudal Europe to its aristocratic roots, because in this first encounter between heavily armoured men-at-arms and mass infantry, the knights lost. That they did so was principally due to the formation the Swiss had developed: the pike phalanx. This was a square formation composed of up to 6000 men, each carrying an 18-foot ash pole with a 4-foot steel spike on the end. The men moved shoulder to shoulder for protection since they wore no armour, and this gave them great mobility. They could move at the run and when necessary change direction backwards or sideways almost instantaneously; they could lower their pikes in any one of four directions, and thus moved swiftly and impregnably across the ground like a giant porcupine. Against this formation there was nothing that men-at-arms could do.

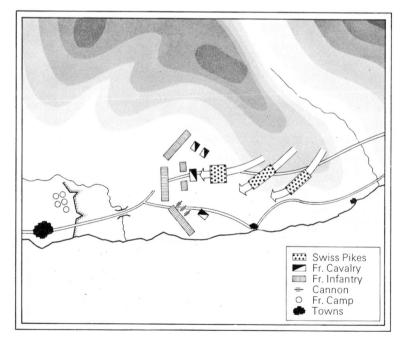

Swiss Pikes
Fr. Cavalry
Fr. Infantry
Cannon
Fr. Camp
Towns

The battle of Grandson, 1476. The Burgundians were defeated by the troops of Berne, Basle, Fribourg and Schwyz; these formed no more than the vanguard of the main Swiss Confederate army, having outstripped their comrades in their eagerness to engage Charles.

A contemporary illustration of the Swiss in action at Grandson. The first pike phalanx has reached the foot of the hill and already repulsed two cavalry charges by Charles's army. The second and third phalanx can be seen approaching down the slope to the left, carrying the standards of the towns they come from.

The pikes would impale the horses, bringing their riders down at the feet of the pikemen who could dispatch the armoured rider with ease. Against the phalanx Charles's cavalry was broken to pieces. The rest of his troops fled so fast that they left behind their gold, silver, jewels, even Charles's throne, and Charles fled with them. The defeat was the talk of Europe.

A year later Charles was dead, killed in yet another military disaster, caused partly by the fact that his borrowed money had not arrived in time to prevent his mercenaries from leaving the field. At his death the Bruges bank was found to have extended credit considerably beyond the original limit, so Charles's borrowing finished Portinari too – he was fired. But the European rulers had learned an important lesson in the art of warfare. The new Swiss method was cheap, since it used only men and pikes. It was also, therefore, standardized. All the phalanx needed was discipline, and any soldier foolhardy enough to break ranks would find himself without the protection of the formation. Above all, the pikeman took little time to train compared with the months needed to train an archer, or the years to become proficient in armour. War suddenly looked cheaper, and the European courts turned eagerly to infantry. As a result armies grew in size, and for the next two hundred years the pikeman was to be the mainstay of any general in the field.

Left: Early guns in the arsenal of the Emperor Maximilian (c. 1505), who was the first ruler to attempt to standardize the size and fire-power of his firearms. These hand culverins were between five and six feet long, and probably date from the middle of the previous century.

The pike squares, as they came to be known, had one crucial limitation – they operated at their best on level ground. This was one of the factors that subscribed to their defeat; the other was the introduction of a new weapon. One of the earliest references to its existence comes from the records of the Italian town of Perugia in 1364. It took the form of a metal tube, mounted on a stick. The tube was filled with a measured quantity of the newly discovered gunpowder, wadded with a stick, firing small stones and balls of lead. The mixture was ignited by setting a match to a small quantity of gunpowder placed in a tiny hole in the side of the barrel next to the main charge. It took two men to handle the new firearm, and its early effect must have been derived more from shock than accuracy. The main problem was that lighting the match at the priming hole made it very difficult to aim – a problem that was rectified by the late fifteenth century with the appearance, probably from Nuremberg in Germany, of the match-lock. This was an S-shaped piece of metal pivoted on the stock so as to bring one end of the S, to which the match cord was attached, in contact with the priming powder in the small hole. The resulting weapon was called an arquebus; it weighed up to 20 lb, had a barrel over $1\frac{1}{2}$ inches in diameter, and fired lead or iron balls 200 yards or more, although it can only have been accurate at a range of about 60 yards. The matchlock was operated by the use of a tricker (trigger), which brought the matchlock down to the priming hole more quickly through the releasing action of a spring.

The arquebus altered the shape of battle. No longer did mounted knights lumber in vain against a sea of pikemen. At the battle of Cerignola in 1503, when the great Spanish commander Gonsalvo de Cordoba met the French, the firearm proved its superiority against the pike square. To begin with, Gonsalvo chose his field with care. It lay at the foot of a vine-covered hill, down which the pike squares of the French had to come to engage his troops. These were drawn up behind a ditch dug at the foot of the hill. Both armies placed their cavalry on the wings, for action when needed. Gonsalvo was out-numbered, but unlike the French one-third of his men had arquebuses. The French advanced down the hill in three pike squares which were hampered by the terrain, and thus were not able to maintain the close formation necessary for defence. Into these advancing masses Gonsalvo's arquebusiers fired a hail of lead and iron, while stakes placed ahead of the ditch stopped the advancing pikemen in the same way as they themselves operated their pikes. The French broke up after several hours of confusion, and at this point Gonsalvo brought out his own pikemen to mop up the survivors. After this victory, Gonsalvo took the logical step of uniting the pike and the arquebus into a square, in which the prime function of the guns was to break up the enemy line so that the pikemen could get into it. The formation adopted bore a striking resemblance to a castle of men, with the pikes forming the main square and the arquebusiers being placed at the corners like towers.

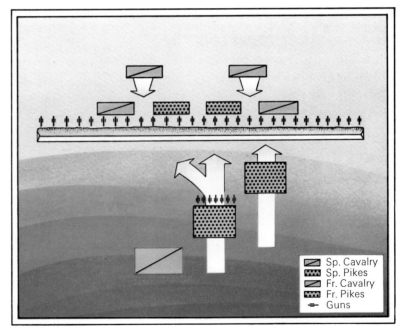

The battle of Cerignola (1503) in south-east Italy – the first successful defence of an en-trenched camp against massed pikemen. Gonsalvo placed his arquebusiers behind the ditch between his artillery. The pikemen behind, and the cavalry on the wings and to the rear, were brought forward only after the guns had disrupted the advancing French pikes.

Sp. Cavalry
Sp. Pikes
Fr. Cavalry
Fr. Pikes
Guns

In 1521 the shape of battle changed again, with the introduction of a new form of arquebus from Spain, called a musket (from the Italian for a sparrowhawk). The new gun was 6 feet long, fired a 2 lb ball, and was heavy enough to demand a forked rest on which to place the gun during firing. By the end of the sixteenth century the weapon was firing balls small enough for there to be twelve to the pound, from which it became known as the '12-bore'. The musketeer's rate of fire was now faster – as many as one shot every four minutes – thanks to the introduction of the bandolier, a belt slung round the shoulder which carried small metal or wooden containers each holding a charge of powder. The soldier would pour the charge of powder into the muzzle, ram in the wad, prime the gun by pouring a small quantity of fine powder on to a ledge set in the side of the barrel next to the hole by the main charge, and pull the trigger. The trigger by now operated a mechanism that looked much like a pecking hen's beak, known as a 'snaphaunce'. This was the immediate forerunner of the flintlock, and in it the trigger released a spring which brought a flint sharply down in contact with a serrated piece of metal, causing sparks which fired the powder in the priming pan.

The Dutch took to this new weapon enthusiastically, and Prince Maurice of Nassau, who was at the time reorganizing the Dutch army, recognized the need for training and discipline in the use of the weapon. For the first time taxes were raised to support a standing army of professional soldiers, obeying published laws at the order of officers and sergeants, operating efficiently because of regular practice in the use of their weapons. One of the earliest books on this new discipline was Jacob de Gheyn's *The Exercise of Arms*, published at the end of the sixteenth century, in which the movements of loading and firing the musket are broken down into forty-two different steps. Maurice of Nassau – and other commanders after him – saw to it that his soldiers were regularly drilled in these movements so that they could be carried out on the battlefield with maximum precision.

Below: Some of the illustrations in de Gheyn's book on musketry. The doffing of the hat indicated to the commander that the musketeer was ready to fire. Note the bandolier of wooden flasks, each one containing a measured charge of powder.

Above: A contemporary engraving of the battle of Breitenfeld, 1631. The Swedish commander's deployment of his musketeers and pikemen can be seen in the left centre of the field, and to the right the cavalry interspersed with musketeers.

This kind of professionalism was to be seen at its finest at the battle of Breitenfeld, four miles outside Leipzig, where on 7 September 1631 the German imperial forces met an army of Swedes and Saxons trained and commanded by the greatest general of the time, the Swedish King Gustavus Adolphus. Gustavus had taken Maurice's ideas to their logical conclusion. He had shortened the musket, making it lighter. He had also shortened the pike from 20 feet to 11 feet, and given his pikemen lighter armour. The shape of the army was changed. The pikemen were now deployed in lines, interspersed with musketeers in the ratio of almost two musketeers to one pikeman, in companies of 126 men. What gave Gustavus the day at Breitenfeld was the fact that he had trained his men to fire in rows of four or five in such a way as to maintain an almost continuous rate of fire. The front man would kneel with a second man standing behind him, and both would fire simultaneously. They would then retire, reloading with meticulous discipline, while the other two came forward to repeat the process. Gustavus also broke with tradition by placing his groups of musketeers between wedges of cavalry, so that from whatever side the enemy came his troops could wheel and fight them off. Against this highly flexible formation the massed squares of the imperial army had no chance. Their cavalry charges failed, and their infantry were devastated by an almost continuous salvo of fire from Gustavus' musketeers, while his pikemen acted as a protective screen to the guns. The armies were now overwhelmingly made up of infantry, whose high rate of fire and greater accuracy gave them increasing power to stop even the strongest cavalry charge.

The shape of war changed yet again in the mid-1700s, with the introduction of the bayonet and the general use of the paper cartridge. The inventor of the bayonet is unknown, although its name probably came from the French town of Bayonne, where there was a flourishing cutlery-making factory. In its early form, it took the shape of a broad blade mounted on a short shaft which plugged into the muzzle. The limitations of this method are obvious: the musket could not be fired while the bayonet was in position. Then in 1690 the ring bayonet first appeared. This was mounted on a ring which fitted round the end of the muzzle and permitted the weapon to be used for both purposes at once. Thus the pikeman was superseded at last, and the new formations in the field gave up the old lines five or ten deep and drew up only two or three deep. The musketeer was now protected against attack while reloading by his own bayonet and that of his comrades. Moreover, the paper cartridge increased the rate of fire. The soldier simply bit off the end, poured the measured charge down the barrel, rammed it home, primed and fired. At the battle of

The battle of Fontenoy, 1745. The painting clearly shows the 'thin red line' of the three-rank British infantry (centre) which kept up such a high rate of fire that it broke charge after charge of the French cavalry. More than 100,000 troops were engaged on the field.

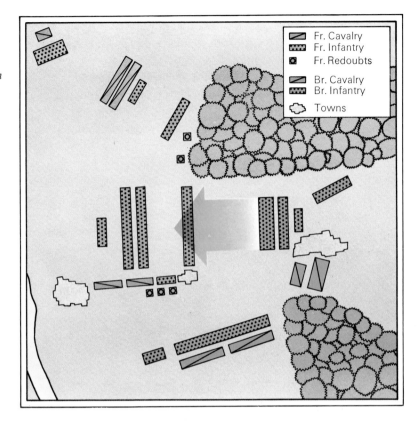

The battle of Fontenoy: the arrow shows the British advance into French ground. As the marching infantry passed between the woods to north and south, they came under crossfire from French redoubts in the woods. The British were eventually driven back by weight of numbers.

Legend:
- Fr. Cavalry
- Fr. Infantry
- Fr. Redoubts
- Br. Cavalry
- Br. Infantry
- Towns

Fontenoy between the Allies and the French, on 9 May 1745, the English wing of the army advanced at a slow steady pace uphill to within thirty yards of the enemy, under severe crossfire from French cannon, then stopped. The front rank knelt, the second took one pace to the right and the third one pace to the left, and from these positions they fired by platoon. The pattern was such that while one platoon was firing the other would reload. By this time the rate of fire of the musket in the hands of an experienced soldier was six times a minute. This devastating volley ripped to shreds the massed French infantry, drawn up in the old-fashioned five-deep line, killing 600 men in the first fire. The French cavalry described action against the English line as like 'charging a flaming fortress'.

All this time, through three centuries of infantry development, the use and scope of cannon grew. By the eighteenth century the siege warfare at which the cannon excelled had greatly slowed down the pace of war. An army might be encamped outside the walls of a city for months, and when it was on the move the horses and wagons needed to transport the guns, ammunition and provisions were numbered in thousands. But it was a political event which radically altered the scale of war.

After the French Revolution, in which the officer corps either fled or were put to death with the other aristocrats, the Directory found themselves with alarmingly few experienced professional soldiers with which to fend off the royalist enemies that ringed France. Their reaction was to take measures which would once again change the shape of battle. They conscripted hundreds of thousands of recruits, to make up six standing armies positioned along France's frontiers from the Channel to the Swiss border. During the early years after the Revolution these armies rarely engaged in formal battle, but when they did the standard formation looked like a return to the long-gone Swiss pike phalanx, since it took the shape of enormous squares of men each with lines of skirmishing musketeers strung out ahead of them. The job of these musketeers was to act as individual marksmen, and to harry the enemy formation from whatever position they chose. They could take cover where they saw fit, advance or retreat as the situation demanded. As soon as their efforts had created maximum confusion in the enemy lines, the giant mass of men behind them would advance and literally roll over the opposing forces by virtue of sheer weight of numbers. The armies on the frontier totalled, according to circumstance and time, anywhere from 600,000 to a million. In 1793 the French Government passed a decree of general levy that in effect introduced the era of mass warfare. The decree said: 'The young men shall fight, the married men shall forge weapons . . . the women will make tents . . . the children shall make up old linen . . . the old men will preach hatred against Kings. The public buildings will be turned into barracks, the public squares into munitions factories.' This was the only expedient open to a revolutionary country, short of trained officers and skilled men. The giant armies were meant to make up in numbers what they lacked in experience and discipline.

Citizens at war. The French levy of 1793 brought the first conscription, as well as involving the population at home in the production of matériel. *The civil involvement continued throughout the Revolution.*

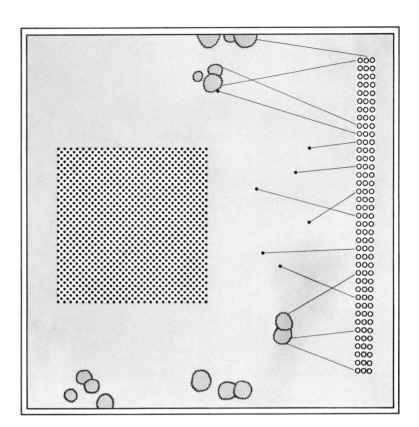

French Revolutionary army tactics in battle. The French infantry mass is on the left, with franc-tireurs *spaced out in front, between them and the enemy. This mass-warfare technique was developed to make up for the lack of officers, exiled or guillotined after the Revolution.*

Early in the life of the frontier armies the problem of provisioning had become acute. Besides the fact that the standing armies had to be fed whether or not their position on the frontier was in an area where food was plentifully available, the general civil structures that had assured adequate food supplies for the population had broken down. Townspeople were doing their own foraging, adding to the dislocation caused by the hungry troops. That problem became acute enough to bring about a major series of changes because it affected Napoleon. He inherited it when he first took command, in 1796, and moved 350,000 men over the Italian border from Nice in order to take Italy from the Austrians. The speed with which he moved an army of that size appears extraordinary until it is remembered that speed is related to supply. Such numbers of men, feeding off the country, must either move on or starve. Napoleon's first campaign in Italy lasted a year, took him all over the northern part of the country, and consisted of a series of victories culminating in a treaty that gave him three papal states, 30 million francs in gold – and a chance of the throne. His astonishing success in battle was due to him being a master of strategy. His answer to the provisioning problem was to split his army into a number of foraging columns, each of which then moved independently and quickly (because of their smaller numbers), remaining near enough to be on call when a battle was imminent while feeding themselves in areas where the entire army could not have been accommodated.

During Napoleon's second Italian campaign this strategy nearly cost him everything, at the battle of Marengo on 14 June 1800. Marengo is today a cluster of four or five buildings at a fork in the road about ten miles outside the city of Alessandria in north-west Italy. Napoleon had arrived with the intention of taking Alessandria from the Austrians; they had come out to meet him, and the two sides met at Marengo. Napoleon spent the night before the battle among his soldiers, watching the Austrian camp-fires to the west. There lay almost the entire Austrian army, over 31,000 men, against Napoleon's reserve army of 28,000. Late that night, Napoleon retired to a villa a few miles away. By six o'clock the next morning the sun was up and it was a fine clear day. The Austrian general, Baron Melas, realizing that most of Napoleon's army was missing – having been sent elsewhere to forage and prepare for later moves in the campaign – suddenly attacked with three columns across the plain, the only one in Italy across which massed cavalry could charge at full speed. They threw themselves at the French, and bitter fighting continued until eleven o'clock when Napoleon arrived from his headquarters, convinced at last that he had a full-scale battle on his hands. By one in the afternoon the French were retreating across the plain to the east. Napoleon sent a desperate message to the nearest column of his foraging troops, to the young general Louis Desaix, saying 'I had thought to attack Melas. He has attacked me first. For God's sake come up if you still can.' When Desaix arrived, at three, the French were still in retreat, reeling under the weight of the Austrian guns, at the village of San Giuliano. Desaix

A panoramic view of the battle of Marengo (1800), showing the death of Desaix (falling from the white horse in front of the trees, left). The Austrian eagle can just be seen on the flags to the right. Note the professional line-formation once more adopted by the French armies; these are no longer made up of Revolutionary conscripts.

charged at the Austrian left flank, and the tide was turned. The battle lasted another hour, finally ending in victory for Napoleon, but at tremendous cost. In spite of the fact that the battle was won by the arrival of Desaix – who lost his own life – at San Giuliano, Napoleon always referred to the day as the battle of Marengo, where he had, in truth, lost.

Marengo was the first victory for Napoleon as Head of State, and it also gave its name to a culinary dish. The night after the battle Napoleon's chef prepared him a dish consisting of a sauce made of oil and water, a chicken, eggs cooked in the sauce with some garlic and tomatoes, and six crayfish, steamed over the chicken as it cooked in brandy. The dish, known as Chicken Marengo, reflected Napoleon's major provisioning problem because, like the food his armies were eating, it was all there was at the time. The French troops could not buy supplies from the local population, who were unwilling to take the *achats*, the paper currency of the revolutionary government. This money was by now practically worthless, inflated as it had been by war with Britain and the Allies and by the economic blockade of France. French industry was being ruined by the war, since supplies of raw materials from overseas were now denied the factories.

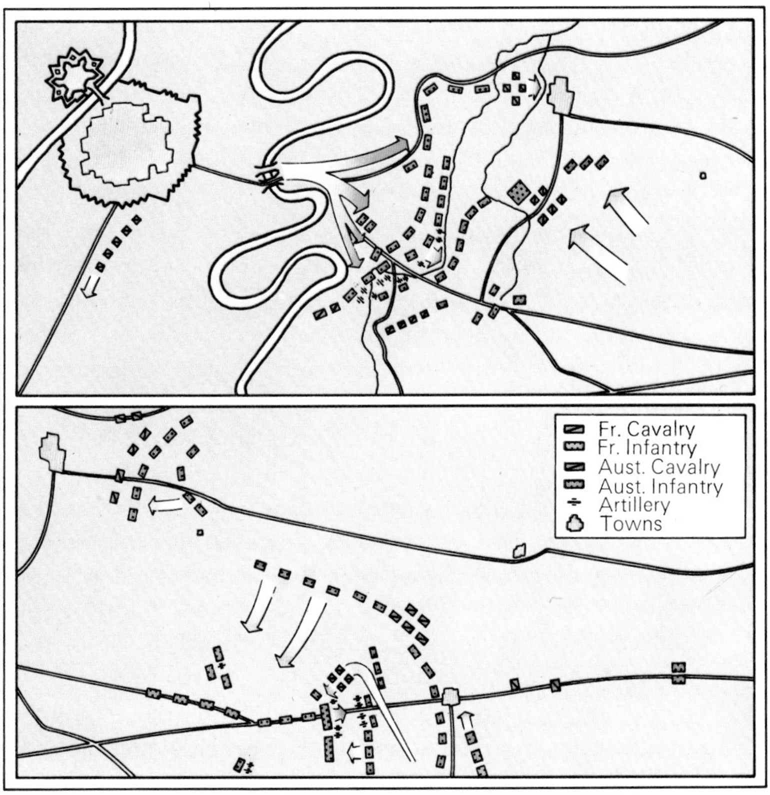

A diagram of the battle of Marengo shows the battle on the morning of 14 June 1800. Melas advances from Alessandria on the left. Marengo sits on the T-junction behind Napoleon's front line.

The afternoon phase is shown below. The French have been driven back (compare the position of the village in the top of the map), when Desaix's charge from the south, crossing the road and wheeling left, saves the day.

Legend:
- Fr. Cavalry
- Fr. Infantry
- Aust. Cavalry
- Aust. Infantry
- Artillery
- Towns

Napoleon saw his provisioning problem as a reflection of the national condition, and decided to rectify it. On his return to France he took the country back to the gold standard, and set about trying to make France economically self-sufficient. The effects of the Revolution and the war had made French industry – never the most efficient on the Continent – underproductive and disorganized. This was the problem Napoleon attacked first. He set up a Society for the Encouragement of Industry, run by a government minister empowered to offer generous prizes for ideas that would help French industry get back on its feet. The prizes went up to 100,000 francs. One of the first people to win a prize was a man who was already fifty when Napoleon was at Marengo. His name was Nicholas Appert. Born in Chalons-sur-Marne of an innkeeper father, Appert had spent his youth working in the wine cellars of the family's hotel in Chalons, and had gone on to become a champagne bottler and cook. By the time he was thirty he had set up a confectionery business in Paris, and then in 1794 he moved a few miles away to Ivry, to ensure fresh supplies of fruit and vegetables for the experiments he was already carrying out. The result of these trials would one day help to solve Napoleon's provisioning problems, win Appert a prize, and lay the groundwork for the appearance of one of the modern world's most useful inventions.

Appert's idea was to preserve food. The container for his first attempts was the champagne bottle. He had handled these bottles often during his earlier years, and, as he said, 'the form of the champagne bottle is most convenient; it is the handsomest, as well as the strongest, and is of the best shape for packing up.' He placed the food to be preserved in the bottles, then sealed them with corks held on, as they were in the wine business, by wire cages. The bottles were then placed in baths of boiling water for varying lengths of time according to the material to be preserved. This ex-cook had inadvertently discovered what no one was to verify until half a century later: that heat sterilized food. The first foods he tried were meat stew, soup, milk, peas, beans, cherries, apricots and raspberries. With the encouragement of the leading gastronome at the time, one Grimod de la Reynière, Appert moved to Massy, about nine miles south of Paris, and set up the world's first bottling factory, employing fifty workers. He sold his bottled foods in Paris at 8, rue Boucher, and soon the shop came to official notice. In 1807 the French Navy took peas, beans and vegetable soup on a trial voyage to the Caribbean, and reported that the results were excellent. The Navy was particularly in need of the kind of independence that bottled provisions could give them, since many of their foreign supply ports were blockaded. In 1809 the newspaper *Courier de l'Europe* took up the story: 'M. Appert has found the art of fixing the seasons. At his hands spring, summer and autumn live in bottles, like those delicate plants which the gardener protects under a dome of glass against the intemperate weather.' By this time Appert's catalogue had lengthened to include asparagus, French beans, artichokes, chicory, spinach, tomatoes, gooseberries, peaches and prunes. On 15 March 1809 the Society for the Encouragement of Industry read a report on some food of Appert's which had been opened after eight months and been found to be in perfect condition, and the following year the society awarded Appert a prize of 12,000 francs on condition that he make his methods known by publishing a book. The work came out in June of that year with the catchy title of *L'art de conserver pendant plusieurs années toutes les substances animales et végétales* (The Art of Preserving Animal and Vegetable Substances). So, partly because the Italians would not accept Napoleon's worthless paper money, Appert gave the world the secret of food preserving. The reason we eat preserved food today mainly from cans and not bottles is, by a curious trick of fate, also because of that money.

Most of the revolutionary currency was made at a paper mill in Corbeil-Essonnes, just outside Paris. This mill was run by a member of the famous French paper-making family called Didot St Leger. The mill had been in existence since 1355, and during the Napoleonic Wars it was experiencing considerable difficulties because many of the skilled men were in the Army, and those remaining were adopting restrictive practices which affected production. In the Paris office of the St Leger firm there happened to be a young clerk called Nicholas Robert, who had a bent for engineering. At some time in the early 1790s Robert was moved to the paper mill to become director of personnel with responsibility for solving the problems of industrial relations plaguing the mill. It was shortly after his arrival that he offered Didot his idea for automating paper production as a means of reducing the number of men needed to run the mill. After Robert had produced a working model, Didot agreed to give him encouragement and financial support to pursue the idea. In 1799 Robert applied for a patent, but due to his limited financial situation was forced to ask for a grant for further work. The government gave him 3000 francs in order to make a fully operational model which would go on public display in the Paris Museum of Arts and Crafts. Arguments soon followed between Robert and Didot, and eventually Didot took over the rights to the invention in return for compensation to Robert. Robert later took him to court and won back the rights, which Didot bought again for a sum which was never paid. Robert left Corbeil, and after a life of trying in vain to exploit his idea, he died penniless. Meanwhile Didot realized that under the circumstances at the time, opportunity for profit from automated paper-making was singularly lacking in war-torn France, and he contacted his brother-in-law.

There followed a chain of events that smacks more of French farce than the sober matter of history as we are so often taught it. Didot's brother-in-law was a man called John Gamble, who worked in the British Royal Navy's prisoner-of-war exchange office in Paris, under its director Captain James Coates. Captain Coates was accorded the privilege of free passage to and from England as part of his duties, and when Gamble mentioned his brother-in-law's plans for taking the Robert invention to England, Coates agreed to help. In 1800 he took Gamble across the Channel to Dover. There he introduced him to the mayor, who invited Gamble to dinner where he met a man by the name of Millikin. Over dinner they discussed Gamble's plans, and Millikin suggested he accompany Gamble to London in order to introduce him to two brothers, Henry and Sealy Fourdrinier, stationers in the City. The Fourdriniers took one look at the plans, and at the long sheet of paper Gamble showed them, and agreed to become involved. In 1801 Gamble obtained a British patent for the machine, and the Fourdriniers set about getting the machines made. Their company millwright was a man called John Hall, who ran a foundry in Dartford, Kent; because he was under pressure of work at the time he passed the paper-mill work on to his mould-maker, Bryan Donkin, who was married to Hall's sister-in-law. In 1803 Donkin set up works at Bermondsey in London, and in the same year erected the first full-scale mill at Frogmore, near Two Waters in Hertfordshire. The main aim was to produce long rolls of wallpaper, in great vogue at the time. The new machines unfortunately did not sell easily, and the Fourdriniers became embroiled in legal action against manufacturers who were infringing the patent. The costs rose alarmingly, and in 1809 the Fourdriniers went bankrupt, leaving Donkin with a works with nothing to make.

Gamble, having helped to get everybody into this mess, cast about for alternative opportunities and, possibly because of his French connections, came in contact with a merchant called Peter Durand, who had recently returned from France with the rights to a patent 'which had been transmitted to him by a foreigner'. The patent was the English version of Appert's original system for preserving food – except that, to cover himself against all eventualities, Durand had added to the original wording: '. . . inclose the said food or articles in bottles or other vessels of glass, pottery, tin, or other metals or fit materials'. It may have been the references to metal that attracted Gamble, and in 1811 he and Donkin and Hall bought the patent rights from Durand for £1000. Appert had himself thought of using metal, but the tin-plate industry in France was practically non-existent, whereas in England it had flourished for over a hundred years. For over twelve months Donkin worked on the idea, and in 1813 sent

Canning food, c. 1870. The limited number of cans which a factory could produce kept the price high. Each can was hand-soldered and contained about twice the amount of an average modern can. The term 'bully beef' comes from this method of preserving food by boiling it (from the French bouillir, *to boil).*

his product to the Royal family. He received a letter in reply which read, 'I am commanded by the Duke of Kent to inform you that his Royal Highness having yesterday procured the introduction of your patent beef on the Duke of York's table where it was tasted by the Queen, the Prince Regent and several distinguished personages and highly approved'. This royal approval was the best advertisement Donkin could have hoped for, and a further letter of approbation followed from Lord Wellesley, President of the Royal Society. The following year the explorer Admiral Ross took Donkin's canned food with him to the Arctic, and in 1815 Otto von Kotzebue did the same on his voyage in search of the Northwest Passage. The partners were made. In 1818 the Royal Navy bought no less than 23,779 cans of meat and vegetables, and several years of profitable monopoly followed. In 1830 the first cans of food reached the shops; they contained tomatoes, peas and sardines. Later the choice of foods increased, but sales went very slowly because of the high price. Soup cost over $7\frac{1}{2}d$, corned beef $8\frac{1}{2}d$ and salmon $11\frac{1}{2}d$ – and this at a time when a family could rent an entire house for $12\frac{1}{2}d$ a week. These early cans were also difficult to open, since doing so involved the use of hammer and chisel! Hand-crafted methods of production also limited output to ten cans a day per man. However, in 1841 the use of chlorine salts in the boiling water raised the temperature enough to complete sterilization in a shorter period of time, and production increased.

Then, in 1845, came trouble. Deaths on the Franklin expedition to the Arctic were blamed on bad food. The cans in question had been produced by a different manufacturer, but the damage was done, and public opinion began to turn against Donkin and others engaged in the trade. The market took a further dive in 1855, when some 5000 cans on their way to the troops in the Crimea were opened and found to be rotten. The suspicion engendered in the public's mind dogged the canning industry until the end of the nineteenth century. At the time it was thought that the cans had let in air, and with it micro-organisms carrying disease. The real reason – that the cans had been insufficiently heated to sterilize the food – was not to be revealed for another decade, when the work of Louis Pasteur became known.

Nearly a hundred years before, an Italian scientist, Lazzaro Spallanzani, who had become a priest in order to earn a living, had boiled up solutions such as gravy for between half and three-quarters of an hour. He then sealed the flasks that contained the solutions, and noted that however long he left the flasks no micro-organisms appeared. He decided that this was because he had killed off all life in the air above the liquid and on the sides of the flask, and had therefore destroyed all the 'spores' that would cause putrefaction. This theory was used to explain the rotten food in the cans at the Crimea, since it was thought that imperfect sealing would have permitted disease-carrying spores to enter the cans. In the same year as the Crimea canning disaster John Gorrie, a doctor working in the American South, died after a lifetime trying to kill the same 'spores' in a different way. His attempts were in the end to contribute to the work on food preservation, although initially his principal concern had been medical.

In the spring of 1833 Gorrie moved to Apalachicola, a small coastal town in Florida situated on the Gulf of Mexico at the mouth of the river after which the town is named. By the time he arrived it was already a thriving cotton port, where ships from the north-east arrived to unload cargoes of supplies, and load up with cotton for the northern factories. Within a year Gorrie was involved with town affairs. In 1834 he was made postmaster, and in 1836 president of the local branch of the Pensacola Bank. In the same year the Apalachicola Company asked him to report on the effects of the climate on the population, with a view to possible expansion of the town. Gorrie recommended drainage of the marshy, low-lying areas that surrounded the town on the grounds that these places gave off a miasma compounded of heat, damp and rotting vegetation which, according to the Spallanzani theory with which every doctor was intimate, carried disease. He suggested that only brick buildings be erected. In 1837 the area enjoyed a cotton boom, and the town population rose to 1500. Cotton bales lined the streets, and in four months 148 ships arrived to unload bricks from Baltimore, granite from Massachusetts, house framing from New York. Gorrie saw that the town was likely to grow as commerce increased, and suggested that there was a need for a hospital. There was already a small medical unit in operation under the auspices of the U.S. Government, and Gorrie was employed there on a part-time basis. Most of his patients were sailors and water-side workers, and most of them had fever, which was endemic in Apalachicola every summer. Gorrie became obsessed with finding a cure to the disease. As early as 1836 he came close to the answer, over sixty years ahead of the rest of the world. In that year he wrote: 'Gauze curtains, though chiefly used to prevent annoyance and suffering from mosquitoes, are thought also to be sifters of the atmosphere and inter-ceptors and decomposers of malaria.' The suggestion that the mosquito was the disease carrier was not to be made until 1881, many years after Gorrie's death, and for the moment he presumed that it came in some form of volatile oil, rising from the swamps and marshes.

Ice farming on the Hudson River in the nineteenth century. Over 100,000 tons of ice left New England every year for the Southern states. There were major ice depots at Charleston, Mobile and New Orleans. Although the ice was cut in winter, warehouse insulation kept it frozen for sale throughout the summer – at progressively rising prices.

By 1838 Gorrie had noticed that malaria seemed to be connected with hot, humid weather, and he set about finding ways to lower the temperature of his patients in summer. He began by hanging bowls full of ice in the wards, and circulating the cool air above them by means of a fan. The trouble was that in Apalachicola ice was hard to come by. Ever since a Massachusetts merchant named Frederic Tudor had hit on the idea of cutting ice from ponds and rivers in winter and storing it in thick-walled warehouses for export to hot countries,

regular ice shipments had left the port of Boston for destinations as far away as Calcutta. But Apalachicola was only a small port which the ships often missed altogether. If the ice crop was poor, the price rose to the exorbitant rate of $1.25 a pound. At some time before 1850 Gorrie had found the answer to the problem. It was well known that compressed gases which are rapidly allowed to expand absorb heat from their surroundings, so Gorrie constructed a steam engine to drive a piston back and forward in a cylinder. The piston first compressed the air in the cylinder, then as it withdrew the air expanded, absorbing heat from a bath of brine in which the cylinder was positioned. On the next cycle the air remained cool, since the brine had given up most of its heat. This air was then pumped out of the cylinder and allowed to circulate in the ward. Gorrie had invented air-conditioning. By bringing the cold brine into contact with water, Gorrie was then able to draw heat from the water to a point where it froze. His first public announcement of this development was made on 14 July 1850 in the Mansion House Hotel, where M. Rosan, the French Consul in Apalachicola, was celebrating Bastille Day with champagne. No ice ship had arrived, so the champagne was to be served warm. At the moment of the toast to the French Republic four servants entered, each carrying a silver tray on which was a block of ice the size of a house-brick, to chill the wine, as one guest put it, 'by American genius'.

Gorrie's air conditioner and freezing machine in action. 1. The piston compresses air, which by (2), at full compression, has become hot. 3. The piston retracts, causing the air to cool down. In (4) the piston returns, to push the cold air out into a cylinder surrounded by a brine-filled jacket. After a number of cycles, the air has extracted all the heat it can from the brine. From then on the air remains cold, and the brine freezes the water in the small container (A).

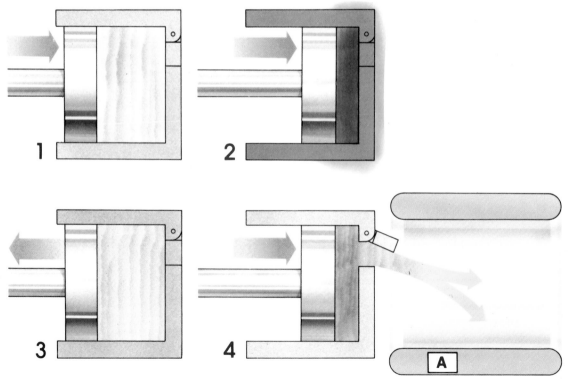

In May of the following year Gorrie obtained a patent for the first ice-making machine. The patent specified that the water container should be placed in the cylinder, for faster freezing. Gorrie was convinced his idea would be a success. The *New York Times* thought differently: 'There is a crank', it said, 'down in Apalachicola, Florida, who claims that he can make ice as good as God Almighty!' In spite of this, Gorrie advertized his invention as 'the first commercial machine to work for ice making and refrigeration'. He must have aroused some interest, for later he was in New Orleans, selling the idea that 'a ton of ice can be made on any part of the Earth for less than $2.00'. But he was unable to find adequate backing, and in 1855 he died, a broken and dispirited man. Three years after his death a Frenchman, Ferdinand Carré, produced a compression ice-making system and claimed it for his own, to the world's acclaim. Carré was a close friend of M. Rosan, whose champagne had been chilled by Gorrie's machine eight years before.

Just before he died, Gorrie wrote an article in which he said: 'The system is equally applicable to ships as well as buildings . . . and might be instrumental in preserving organic matter an indefinite period of time.' The words were prophetic, because twelve years later Dr Henry P. Howard, a native of San Antonio, used the air-chilling system aboard the steamship *Agnes* to transport a consignment of frozen beef from Indianola, Texas, along the Gulf of Mexico to the very city where Gorrie had tried and failed to get financial backing for his idea. On the morning of Saturday 10 June 1869 the *Agnes* arrived in New Orleans with her frozen cargo. There it was served in hospitals and at celebratory banquets in hotels and restaurants. The New Orleans *Times Picayune* wrote: '[The apparatus] virtually annihilates space and laughs at the lapse of time; for the Boston merchant may have a fresh juicy beefsteak from the rich pastures of Texas for dinner, and for dessert feast on the delicate, luscious but perishable fruits of the Indies.'

At the same time that Howard was putting his cooling equipment into the *Agnes*, committees in England were advising the government that mass starvation was likely in Britain because for the first time the country could no longer feed itself. Between 1860 and 1870 consumption of food increased by a staggering 25 per cent. As the population went on rising, desperate speeches were made about the end of democracy and nation-wide anarchy if the Australians did not begin immediately to find a way of sending their sheep in the form of meat instead of tallow and wool. Two Britons, Thomas Mort and James Harrison, emigrated to Australia and set up systems to refrigerate meat. In 1873 Harrison gave a public banquet of meat that had been frozen by his ice factory, to celebrate the departure of the S.S. *Norfolk* for England. On board were twenty tons of mutton and beef kept cold by a mixture of ice and salt. On the way, the system developed a leak and the cargo was ruined. Harrison left the freezing business. Mort then tried a different system, using ammonia as the coolant. He too gave a frozen meat lunch, in 1875, to mark the departure for England of the S.S. *Northam*. Another leak ruined this second cargo, and Mort retired from the business. But both men had left behind them working refrigeration plants in Australia. The only problem was to find the right system to survive the long voyage to London. Eventually, the shippers went back to Gorrie's 'dry air' system, and the first meat cargo to be chilled in this fashion left Australia aboard the S.S. *Strathleven* on 6 December 1879 to dock in London on 2 February of the following year with her cargo intact. The meat sold at Smithfield market for between 5*d* and 6*d* per pound, and was an instant success. Queen Victoria, presented with a leg of lamb from the same consignment, pronounced it excellent. England was saved.

Harrison's first attempts at refrigeration in Australia had been in a brewery, where he had been trying to chill beer, and although this operation was a moderate success, the profits to be made from cool beer were overshadowed by the immense potential of the frozen meat market. The new refrigeration techniques were to become a boon to German brewers, but in Britain, where people drank their beer almost at room temperature, there was no interest in chilling it. The reason British beer-drinkers take their beer 'warm' goes back to the methods used to make the beer. In Britain it is produced by a method using a yeast which ferments on the surface of the beer vat over a period of five to seven days, when the ideal ambient temperature-range is from 60 to 70°F. Beer brewed in this way suffers less from temperature

Right: Unloading frozen meat from Sydney in the South West India Dock, London. This shows the hold of the Catania, *which left port in August 1881 with 120 tons of meat from the same exporters who had filled the* Strathleven.

Below: The S.S. Strathleven, *carrying the first successful consignment of chilled beef from Australia to England. Note the cautious mixture of steam and sail, which was to continue into the twentieth century.*

changes while it is being stored, and besides, Britain rarely experiences wide fluctuations in summer temperatures. But in Germany beer is produced by a yeast which ferments on the *bottom* of the vat. This type of fermentation may have been introduced by monks in Bavaria as early as 1420, and initially was an activity limited to the winter months, since bottom fermentation takes place over a period of up to twelve weeks, in an ideal ambient temperature of just above freezing point. During this time the beer was stored in cold cellars, and from this practice came the name of the beer: lager, from the German verb *lagern* (to store). From the beginning there had been legislation in Germany to prevent the brewing of beer in the summer months, since the higher temperatures were likely to cause the production of bad beer. By the middle of the nineteenth century every medium-sized Bavarian brewery was using steam power, and when the use of the piston to compress gas and cool it became generally known, the president of the German Brewer's Union, Gabriel Sedlmayr of the Munich Spätenbrau brewery, asked a friend of his called Carl von Linde if he could develop a refrigerating system to keep the beer cool enough to permit brewing all the year round. Von Linde, an engineer and an academic, solved Sedlmayr's problem, and gave the world an invention that today is found in almost every kitchen.

243

Von Linde used ammonia instead of air as his coolant, because ammonia behaved just as air did: it liquefied under pressure, and when the pressure was released it returned to gaseous form, and in so doing drew heat from its surroundings. In order to compress and release the ammonia, he used Gorrie's system of a piston in a cylinder. Von Linde did not invent the ammonia refrigeration system, but he was the first to make it work. In 1879 he left his university work and set up laboratories in Wiesbaden to continue research, and to convert his industrial refrigeration unit into one for the domestic market. By 1891 he had put 12,000 domestic refrigerators into German and American homes. The modern fridge uses essentially the same system as the one with which von Linde chilled the Spätenbrau cellars.

Interest in refrigeration spread to other industries. The use of limelight, for instance, demanded large amounts of oxygen, which could be more easily handled and transported in liquid form. The new Bessemer steel-making process used oxygen. It may be no coincidence that an ironmaster was involved in the first successful attempt to liquefy the gas. His name was Louis Paul Cailletet, and together with a Swiss engineer, Raoul Pictet, he produced a small amount of the liquid in 1877. All over Europe scientists worked to produce a system that would operate to make liquid gas on an industrial scale. One such was extremely efficient. It worked because of what was known as the Joule–Thomson effect, and liquefied gas by repeatedly releasing it through a very fine jet, so that its pressure fell rapidly at each stage.

The major problem in all this was to prevent the material from drawing heat from its surroundings. In 1882 a French physicist called Jules Violle wrote to the French Academy to say that he had worked out a way of isolating the liquid gas from its surroundings through the use of a vacuum. It had been known for some time that vacua would not transmit heat, and Violle's arrangement was to use a double-walled glass vessel with a vacuum in the space between the walls. Violle has been forgotten, his place taken by a Scotsman who was to do the same thing, much more efficiently, eight years later. His name was Sir James Dewar, and he added to the vessel by silvering it both inside and out (Violle had only silvered the exterior), in order to prevent radiation of heat either into or out of the vessel. This meant that it could equally well retain heat as cold. Dewar's vessel became known in scientific circles as the Dewar flask; with it, he was able to use already liquid gases to enhance the chill during the liquefaction of gases whose liquefaction temperature was lower than that of the surrounding liquid. In this way, in 1891, he succeeded for the first time in liquefying hydrogen. By 1902 a German called Reinhold Burger, whom Dewar had met when visiting Germany to get his vessels made, was marketing them under the name of Thermos.

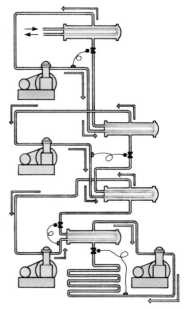

Above: The cascade refrigeration system. At each of the four stages the technique of compression and expansion is used to refrigerate a gas to the liquid state. It then expands, drawing heat from another gas, chilling it down so that it can be liquefied at an even lower temperature so as to help refrigerate the next gas, and so on, until the last stage. This involves liquefying the gas which demands the lowest temperature possible.

Right: Sir James Dewar lecturing at the Royal Institution. Although Violle preceded Dewar in the development of the vacuum flask, there is no evidence that Dewar knew of his work when he presented the details of his new container to the Royal Institution in 1890.

The impact of the vacuum flask was wide and varied. Initially it changed the social habits of the Edwardian well-to-do: picnics became fashionable because of it. In time it changed the working-man's lunchbreak, and accompanied expeditions to the tropics and to the poles, carrying sustenance for the explorers and returning with hot or cold specimens. Later, it saved thousands of lives by keeping insulin and other drugs from going bad. Perhaps its most spectacular impact was made, however, by two men whose work went largely ignored, and by a third who did his work in a way that could not be ignored. The first was a Russian called Konstantin Tsiolkovsky, whose early use of liquid gas at the beginning of the twentieth century was to lie buried under governmental lack of interest for decades. The second was an American called Robert Goddard, who did most of his experiments in the 1930s on his aunt's farm in Massachusetts, and whose only reward was lukewarm interest from the weather bureau. The third was a German, Herman Oberth, and his work was noticed because it aimed at the destruction of London.

Being able to keep things cold or hot greatly enhanced the popularity of picnics – enthusiasts such as these could now drink hot soup at the coldest outdoor meeting. The photograph shows spectators at the Bath Horse Show in September 1910.

The V-2 liquid-fuelled rocket used a mixture of oxygen and kerosene. Originally developed at the experimental rocket base in Peenemunde, on the Baltic, the first V-2 landed in London in 1944. When the war ended, German engineers were working on a V-3 capable of reaching New York.

His liquid gases were contained in a machine that became known as Vengeance Weapon 2, or V-2, and by the end of the Second World War it had killed or injured thousands of military and civilian personnel. All three men had realized that certain gases burn explosively, in particular hydrogen and oxygen, and that, since in their liquid form they occupy less space than as a gas (hydrogen does this by a factor of 790) they were an ideal fuel. Thanks to the Dewar flask they could be stored indefinitely, transported without loss, and contained in a launch vehicle that was essentially a vacuum flask with pumps, navigation systems, a combustion chamber and a warhead.

One of Oberth's most brilliant assistants was a young man called Werner von Braun, and it was he who brought the use of liquid fuel to its most spectacular expression when his brainchild, the Saturn V, lifted off at Cape Canaveral on 16 July 1969, carrying Armstrong, Aldrin and Collins to their historic landing on the moon.

9
Lighting the Way

On 20 July 1969, at 13 hours, 19 minutes, 39.9 seconds U.S. Eastern Standard Time, the front and left legs of the lunar module Eagle hit and skidded three feet in the grey dust of the Sea of Tranquillity, and the module rocked to a halt. Man had landed on the moon at the end of a journey lasting 4 days, 6 hours, 45 minutes and 39.9 seconds. The accuracy of these timings reveals the navigational precision with which the flight was undertaken, since the only real reason for needing to know what time it is during a journey is in order to find out where you are. In the case of Apollo 11, the length of journey, complexity of trajectory and potential hazards made precise knowledge of the spacecraft's position particularly vital. Getting Armstrong, Aldrin and Collins to the moon was not merely a matter of pointing the rocket into the sky and firing the engines. The launchpad itself was moving in space, as the Earth turned at over 1600 m.p.h. At the time of launch Cape Canaveral was on the opposite side of the Earth from the moon, which was itself circling the Earth. Finally, the target in space for insertion into orbit round the moon was a point 60 miles above the lunar surface, although at the time of launch the moon had not yet arrived at that point in space. The extraordinary fact is that when the lunar module came in over the target area, over 240,000 miles from Earth, mission rules demanded that with no further alterations to its trajectory than those already in the computer, it should be no more than one and a half miles off centre. All six Apollo landings between 1969 and 1972 were achieved with this degree of accuracy, thanks to the inertial navigation system aboard the spacecraft.

The Apollo Saturn V launch vehicle at lift-off. The stack consisted of three separable sections, each of which housed fuel tanks and rocket motors. The third, top section fired to put the spacecraft on translunar trajectory.

This system was part of the Guidance and Navigation Control unit, in which navigational information was used to fire the engines at points in the flight where changes in direction were required. Knowing when to perform these firings depended on knowing where the Apollo was in space, and this information was provided by an accelerometer and a clock. The accelerometer consists of a container filled with extremely dense liquid surrounding a piece of metal only slightly denser than the liquid itself. If left free, the metal will tend to sink very slowly to the bottom of the container. Since there is no gravitational attraction in space such as there is on Earth, some alternative method has to be found to maintain the metal in a stable position. This is done by the use of electromagnets surrounding the container. The magnetic field of these magnets will hold the metal at the exact centre of the liquid with the minimum amount of necessary power.

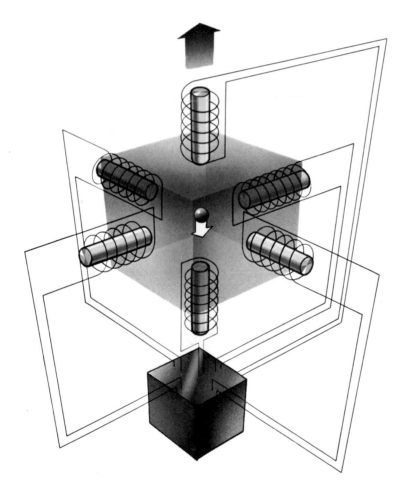

The accelerometer senses movement in six directions: pitch (vertical displacement up and down), roll (spin in either direction about the directional axis) and yaw (lateral displacement to one side or the other of the axis). In a modern jet aircraft, any inertial guidance system not programmed with a destination will fly the plane to 0° 0" latitude and longitude – the point where the Greenwich meridian intersects the equator, off the west coast of Africa!

Thus, if the container moves in any direction, the metal will drift a very small amount in the opposite sense, in reaction to the movement. The amount by which the metal moves will relate to the force with which the container moves, i.e. its acceleration. The movement of the metal causes changes in the magnetic field set up by the magnets, which are programmed to return to field to its original state. In so doing they produce enough extra power in the right area to move the metal back to its original position, thus restoring the status quo. In order to do this, magnetic power is increased to push the metal back in the direction it came from, which is the direction in which the container is moving. If the container is attached to the spacecraft, the direction in which the metal piece is moved and the power needed to move it show the course and acceleration of the spacecraft. This checking and repositioning of the metal piece is done many times a second. All that is necessary in order to establish the position of the craft in space is to know in what direction acceleration has been taking place, and for how long. It is for this reason that the Apollos were fitted with clocks accurate to within one-hundredth of a second.

The Guidance and Navigation System is a perfect example of why science and technology produce double-edged weapons. On the one hand, inertial navigation can be adapted easily to move the control surfaces of a modern jet aircraft at predetermined times, and to alter the power setting of the engines so that an airliner can fly to its destination and land without more than monitoring assistance from its crew. On the other hand, the same technology contributed to the dangerous strategic imbalance that existed during the cold war standoff. Developments then in inertial navigation radically altered the accuracy of intercontinental ballistic missiles. They carried multiple warheads, each one of which could be separately and independently aimed at its destination. Each of those warheads (called Multiple Independently-targetted Re-entry Vehicles) could strike within one hundred yards of its aim point.

It was this precision of attack that upset the balance. Missiles which had earlier been secure, either in concrete silos which could survive atomic explosions only a few miles away or in submarines which could not be detected, were no longer safe. They would not be likely to escape destruction from incoming MIRVs. In this case they could not be used for a second, retaliatory strike after the initial enemy attack. This second strike capability – the chance that enough missiles would survive a pre-emptive attack to cause unacceptable damage to the attacker – used to be the deterrent against carrying out the first strike. With a minimal chance that the missiles would survive in order to carry out a second strike, their effective use was now limited to the aggressive role. The question was, who would decide to press the button first?

252

It was just such a major breakthrough in missile technology which necessitated a similar reappraisal of warfare and defence over five hundred years ago, when the cannon was invented. Until the arrival of gunpowder, well-built castles were virtually impregnable. They consisted of a central, square keep to which the owner could retire should his mercenaries revolt, and which could also be a last-ditch defence should the enemy penetrate the walls. These keeps were surrounded by a deep ditch crossed at the main entrance by a removable drawbridge. The castle often had more than one perimeter wall; these were built tall and thin, and for this reason were called curtain walls. The purpose of high walls was to make them difficult to scale: the longer it took the enemy to climb, the greater the chance of killing him before he reached the top. Height meant protection also against stone-throwing catapults, and against wooden siege engines which attempted to lift men as high as the top of the wall so that they could fight the defenders on their own level. Along the walls ran crenellations (battlements) and machicolations, which were structures jutting out from the wall from which oil and debris could be dropped on the heads of the enemy.

One of the most perfect examples of pre-gunpowder fortification is the walled city of Aigues Mortes, on the Mediterranean coast of France. This city was begun by Saint Louis, who needed a port from which to launch his crusade to the Holy Land, and finished by his son Philip in 1300. The entire town and fortifications were designed as a piece, and colonists were attracted to live there by the promise of tax concessions and certain legal rights – such as, in the case of women, freedom from prosecution for adultery. The chequer-board street plan was designed so that troops could be moved with maximum efficiency from one part of the walls to another during siege. Aigues Mortes and other cities and castles like it survived with impunity the attacks of armoured knights, bowmen, infantry and catapults.

With the arrival of the cannon they were all rendered instantly obsolete. The high curtain walls, so long a secure defence, were easy targets. At the beginning of the fifteenth century various tactics were adopted to protect the castles from the relatively inaccurate bombardment of the early cannon, which hurled piles of stones and rubble. The walls were thickened, sometimes up to fifty feet, and were covered with timber or earth to withstand impact. The crenellations and machicolations were removed, since they were easily knocked off, to fall and crush the defenders behind the walls. Moats were widened so that falling masonry could no longer fill them and form bridges for the attackers. These modifications worked relatively well until developments in metallurgical techniques, coupled with better quality gunpowder, led in the middle of the fifteenth century to the use of more

The siege of Constantinople by the Turks in 1453. Before the development of efficient cannon, the usual method adopted was to blockade supplies to a city (here the blockade is effected by the Turkish ships in the background) and hope to starve the defenders into submission.

accurate guns which fired iron cannon-balls. The effect of the new guns was devastating, both physically and psychologically. Towns and castles would surrender at the very mention of their arrival.

Throughout the fifteenth century, however, defences went on being developed against the invincible ball. The first casualty, in spite of the advice of Machiavelli that they should be retained because they were difficult to scale, was the high curtain walls. Although these walls were indeed difficult to scale, they were also very easy to hit with a cannon-ball; so they were lowered, to provide a more difficult target. Earlier it had been the practice to build the base of the old round towers out at an angle – a design known as 'scarping' – so as to lessen the impact of battering rams. This proved equally effective against the cannon, since the angle of the masonry caused the ball to ricochet, thus losing much of its impact. Walls and towers began to be constructed in brick, which was more capable of absorbing that impact than stone. As the fifteenth century drew to a close, interest in scientific theory and in mathematics particularly was stimulated by the translation and dissemination through printed books of previously unknown Greek texts. As these works were initially produced in Italy it is not surprising that the application of the new mathematics to architecture and fortification should have come from Italy. It was there that the first of the new style of defensive structure appeared.

Besieging a castle in the late fifteenth century. Note the soldier (bottom right) lighting the charge on the new terror weapon, the bombard. These guns could be fired, at most, ten times a day. Here temporary forward defences of timber and earthworks have been set up in an attempt to lessen the effects of the shot.

In its early form it solved one major problem that the medieval fortifications faced with the advent of the cannon. At the foot of the old round tower lay a considerable area of dead ground, visible only to those standing at its outermost point, where they could look straight down. The field of view from either side of the tower left a triangular-shaped area of ground that could not be seen. Initially this area was used to advantage by attackers who could creep right up to the base of the tower and lay mines. This strategy often led to the creation of a veritable rabbit-warren underneath the walls, as the miners and counter-miners went to work. The miners would lay their charges, and the counter-miners would try to dig close enough to undermine the walls of their tunnels. The problem became such that the great Italian engineer and architect Leon Battista Alberti recommended a system of pipes running through a citadel and its foundations to carry the miners' conversations to the listening defenders. The final answer to the triangular area of dead ground existing beyond the outermost point of the round tower seems obvious – it was filled in by the construction of a tower that was itself triangular in shape. One of the earliest Italian towns to be protected by this new type of tower, known as the 'bastion', was the Tuscan city of Arezzo, whose bastioned

Above: Greek fire being used by defenders against the attacking ships. This was a burning mixture of oil, sulphur and pitch which was first used by the Byzantines against the Arabs, c. 675. Although sometimes projected by crude force pumps, its range was far more limited than that of the earliest cannon.

Right: The defenders strike back at the beginning of the seventeenth century, by placing cannon on the castle walls. Now it is the turn of the attackers to defend themselves, using baskets full of earth to protect their guns. Note their munitions store – a barrel of gunpowder and spare cannon-balls.

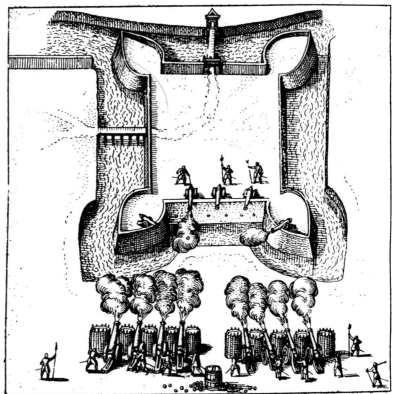

The fully developed star fort of the seventeenth century, showing bastions, ravelins and moat. The fortress is protected from every conceivable angle. Note the town streets, radiating from a central open space, where the captain of the fort set up his command post.

walls were constructed in 1503 by Giuliano di Sangallo. The finest extant example of the new fortress design is the Fortezza da Basso in Florence, designed by Antonio di Sangallo and hailed at its completion in 1533 as impregnable, the model for Europe. This was a hundred years after the point had sunk home. Alberti had written, some time after 1440: 'If you were to look into the expeditions that have been undertaken, you would probably find that most of the victories were gained by the art and skill of the architects rather than by the conduct or fortune of the generals, and that the enemy was more often overcome and conquered by the architect's wit without the captain's arms, than by the captain's arms without the architect's wit.'

The new design permitted a clear field of vision over every inch of wall, since the jutting sides of the triangle were themselves built along a line which was a continuation of the angle of vision available to the gun positions on the walls on either side of the tower. By the middle of the sixteenth century bastions were being built all round the walls of fortresses and fortified cities. Apart from topographical considerations which affected their positioning, the bastions were usually placed at intervals which corresponded to the range of the guns placed on each bastion, the point being that one bastion could defend another from attack. This factor limited the distance between one bastion and another to approximately 300 yards; it also limited the angle of the bastions' sides, which had to correspond to the line of sight of guns on the adjacent bastions, since it was part of the task of such guns to cover the sides of neighbouring bastions with enfilading fire. This left one major problem unresolved: if the attacking forces reached the main wall, gunners on the bastions were forced to fire back at their own defences in order to hit the enemy. A refinement in bastion design was introduced later in the century to counter the difficulty. 'Oreillons' were set into the part of the bastion that joined the main wall, so as to give the bastion an arrowhead shape. From

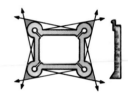

The problem of dead ground in front of drum towers, not covered by lines of fire, and the thin curtain walls between the towers.

Early attempts to solve the problem. The angled brick wall caused less damaging ricochets.

The final bastion principle, in simple form.

256

these oreillons the defenders could fire at the enemy along the wall, instead of directly at it. In order to keep the enemy guns as far away as possible, the ideal fortification was surrounded by a moat in which, at points equidistant between the bastions, small islands were built. These islands, called 'ravelins', were triangular in shape and served as a defensive outpost from which crossfire could be set up against any attempts to cross the moat. Should one of the ravelins be captured, they would be vulnerable to crossfire from the bastions. With a defensive wall structured in this manner, towns within the circuit of walls were rebuilt so that their streets radiated out from the town centre to each bastion, and along behind the walls so that cannons and munitions could be moved with maximum ease from one defensive point to another during periods of siege. The final shape of the new defensive structures resembled a star, and for this reason they were known as star forts. They were to be the principal means of defence throughout Europe and across the world from the Caribbean to India until late in the eighteenth century. The engineers and architects who built them carried sample books which they would show to the authorities in a town, and a choice was then made on the basis of how many bastions the town could afford. Many of these star forts have been destroyed or partly obliterated by later civil expansion; one of the most perfect surviving examples is to be seen at Naarden, in Holland, where the fort remains virtually untouched.

A contemporary painting of the fully fortified town of Gravelines, near Dunkirk. At this time (1652) it was occupied by the Spanish, but six years later it was captured by the French.

These major changes in architectural style stemmed, as has been said, directly from the arrival of the cannon. The surveying and measuring instruments that were used to design and build them were also military in origin, developed from earlier astronomical instruments, for the most part in order to increase the accuracy of guns. For the first two hundred years of the existence of the cannon each gun's performance tended to be unique, and was measured by a process called 'zeroing'. The cannon was first fired level, and note was made of where the ball fell. Then the gun was elevated, one degree at a time, fired, and the fall of the shot noted. In this way a profile was drawn up of the gun's performance. The gunner knew how far his gun would fire at certain angles, and his responsibility for working out what such angles were was relieved in 1537 by an Italian engineer called Nicola Fontana (nicknamed Tartaglia, the Stammerer).

His device for measuring angles was called a gunner's quadrant. It showed the angle at which the gun was elevated, and later versions carried scales with which to measure the charge of powder accurately, so as to make sure that the force of the explosion was consistent each time. The question then was to measure how high and how far away the target was, in order to elevate the gun to the correct degree for that range. At this time height was measured with an astronomical

Gunners calculating the elevation of their artillery. The man on the right is working out the height of the target, while his colleague uses a quadrant to set the cannon. The plumb line shows the angle of elevation which is altered by inserting or removing wooden pegs and wedges under the barrel.

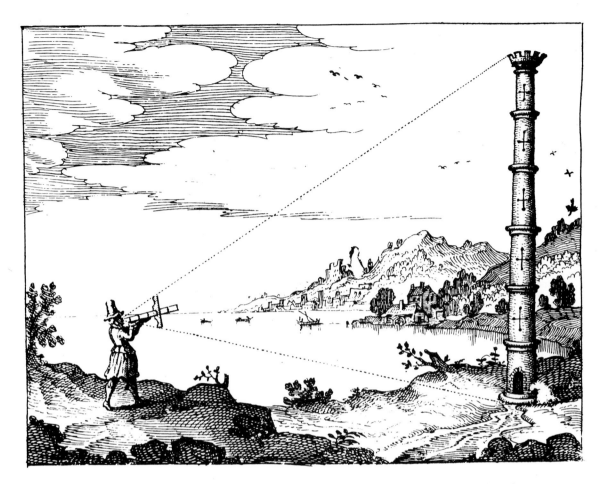

A form of cross-staff being used to
calculate the height of a building.
The observer moves the vertical
section of the staff to align the top
and bottom with the summit and
base of the building. The angles
are then read off a graduated scale
along the horizontal section.

instrument called the cross-staff. It had been introduced to Western
Europe by the great Jewish translator Levi ben Gerson during his
fourteenth-century work on Arab texts. In 1544 the technique of
using the cross-staff became widely available when the Dutch
astronomer Gemma Frisius, who was Professor of Mathematics at
Louvain University, published what was to become the standard work
on the subject.

Frisius had also provided the gunners with the answer to their other
problem – that of knowing how far away the target was – in an earlier
book published in Antwerp in 1533, in which he had described the
concept of triangulation for the first time. This involved one
of the last uses of the astrolabe, before it was replaced by the new instruments of the seventeenth century. Frisius placed the astrolabe on its
side, and replaced its expensive brass construction with wood. The
sighting device was now merely a pointer with sights or pins. This
disc and sight were to be placed flat, and the 'nautical instrument which
we call a compass' was placed on it so that the edge of the compass-
box sat along the disc's zero line, i.e. the north–south line. The entire

instrument was then rotated until the compass was pointing towards magnetic north on the compass rose. The compass could now be lifted off the disc, as long as the latter were fixed in place. With the disc oriented so that its north–south line lay from north to south, the angle at which any target stood to the observer (or gunner) could now be read off by swivelling the sights just as on the astrolabe, except that in this case of course the angle being read was along the horizontal plane. Once this angle was established the gunner moved the whole set of instruments to another position from which he could see the target, and repeated the process. Having measured the distance between his two observations, he now had the data with which to work out the position of the target. He had a baseline of known length, and two sides of a triangle, each of which represented the line drawn between one of his observation points and the target. These lines intersected at the apex of the triangle. All he had to do was draw the triangle to scale, and then measure the distance between the position of his gun on the triangle base, and that of the target at the intersecting point, and the result was the distance his gun had to fire. Frisius, whose ideas were passed on to his most brilliant pupil, Gerhardus Mercator, had enabled ranging on the battlefield to be carried out easily and quickly, and in consequence his theories rapidly became popular. By 1550 they were accepted virtually everywhere.

The only drawback to the new system was that it needed three separate sets of instruments: the astrolabe, the planimetrum (the wooden astrolabe disc) and compass, and the means to draw the triangle with accuracy. This last was called a plane-table. It began life as a drum head, on which the gunners drew out their triangles by rule of thumb, and developed into the sophisticated instrument produced in 1551 by the French courtier Abel Foullon. It consisted of a table mounted on a pedestal by means of a ball and socket joint, which allowed the table to rotate freely. A compass was set in the centre of the table with the degrees of the circle marked around it on the table. Along one edge of the table was a scale marked in a thousand parts, and pivoted on each end of this scale was a ruler. On each side of the table a semicircle of brass with degrees marked on it was mounted on a vertical rod, and carried a sight. With this plane-table all measurements can be made. The semicircles at the side measure altitude (degrees above the gun), while the rest of the table gives azimuth (degrees to either side of the gun). Although Foullon's new plane-table put all the parts together in one instrument, it was a complex and cumbersome device to carry to battle and time-consuming in use, especially during siege action when the enemy in the fortress would already be familiar with the range and distance immediately round the fortifications. Nor was the plane-table easy for illiterate artillerymen to master.

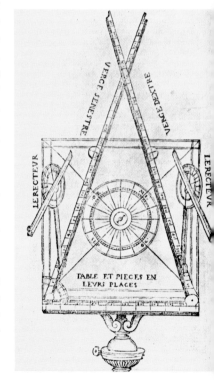

Foullon's plane-table (or holometer) – the first attempt to enable gunners to make all their calculations with one instrument. Once the table was set up with the aid of the compass, further calculations were made on a sheet of paper fixed on to the table with blobs of wax.

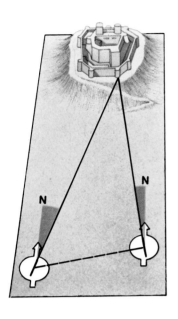

Digges' theodelitus, the first efficient surveying instrument that was also easy to use. A compass was used to set it up, as illustrated by the diagram, above left. Wherever the operator moved, angle measurements could be made accurately from a constant reference point (i.e. north).

In 1551 another, more simple variant of the same instrument was produced by Sebastian Münster, who replaced the degree scale with a simple division by numbers, and provided a reference table which did all the mathematics for the gunners. Then in 1571 a book was published in England called *A Geometricall Practise, named Pantometria* by Leonard Digges, one of the leading mathematicians of Elizabeth's time. In this book Digges described his new *theodelitus*, and brought together all the knowledge of the period regarding the measurement of height and distance. The instrument did all that the Foullon plane-table did, but more efficiently and, most important, more quickly. There was one graduated semicircle instead of two, and it was mounted directly above the horizontal circle on which degrees of azimuth were marked. The same sight served for both azimuth and altitude simultaneously. The operator merely pivoted the altitude semicircle on its vertical support, lined up the sights on the target, and in this way was able to read off the degree of altitude on the semi-circle and, through a pointer fixed to the vertical support, the degree of azimuth on the horizontal circle. The theodelitus also had a compass built in to its horizontal circle and a plumb-line hanging from the semicircle, so that the instrument could be quickly and easily aligned in both vertical and horizontal axes whenever it was moved. The military was satisfied.

In 1590 Sir William Segar, in his *Book of Honor and Armes*, said: 'Who without learning can conceive the ordering and disposing of men in marching, encamping, or fighting without arithmetic?' He might well have asked, because the business of moving from one place to another was still very much an affair of chance, due to the poor quality of the maps available. Even in the middle of the sixteenth century, with the sophisticated measurement and design involved in building bastions, the armies that marched to put them to siege used maps that had not changed much in the previous three hundred years. They consisted of little more than a number of wavy lines denoting rivers, and the names of the towns and villages where bridges had been built to cross them. By Digges' time, however, these maps were improving, thanks to the general improvement in communications and the economy of England, and not least to Henry VIII's divorce proceedings.

After disagreement with the Pope over his wish to divorce Katherine of Aragon, Henry had dissolved the monasteries in 1536 and confiscated their lands. Apart from his desire to get rid of his first wife, Henry realized that the monastic holdings would provide the money he needed to finance both the fortification of the south coast of England and the wars with the Scots and the French. The amount of land in question was considerable. Exactly how much, nobody knew. In 1530 a survey had shown that there were no less than 502 monasteries, 136 nunneries and 187 friaries in the country, with an estimated population of over 9000. Well before the Dissolution, Henry was preparing the ground. Between autumn 1535 and spring 1536 Thomas Cromwell had sent 'visitors' to report on life in the religious houses. How detailed their survey was may be judged by the fact that they claimed to have visited 120 northern monasteries in two months. Considering the condition of the roads at the time, this accomplishment was well nigh impossible. But they brought back what the king wanted, in the form of reports of 'manifest sin, vicious, carnal and abominable living'. Their brief reveals the purpose for which they were sent, since they were asked to find out only if there were sex offences being committed, the names of those who wanted to leave a community, what valuable relics were present, the name of the descendant of the founder, and most important of all, the income and debts of the house.

The Dissolution is often pictured as a wave of terror. In mitigation it should be said that although all the lands were taken by the crown, monastic debts were honoured, and monks and nuns were permitted to go to houses which had not been dissolved, or were given pensions. By 1538 the monasteries were surrendering at the rate of twenty a month. One which did not surrender was Glastonbury, in Somerset, whose abbot was called Richard Whiting. Eventually evidence was brought to show that the abbot had indulged in embezzlement of monastic funds, and he was hanged, drawn and quartered on Glastonbury Tor for this and other alleged crimes. His steward happened to be a man called Jack Horner, and if the malicious local poem is to be believed, he did well out of the abbot's demise. The abbot had apparently tried to appease Henry with a gift of twelve title-deeds to land. This offer failed, but not before Horner had taken one of them out of the piecrust in which they had been placed for effect on presentation. The 'plum' that Horner is supposed to have pulled out was a title-deed to land which he and his family then enjoyed. The surveys, aided by informants such as Horner, revealed that the monasteries had more money than did the crown. They were quickly relieved of it. Lead was stripped from the roofs, church bells were melted for cannon, their gold, silver and jewels were sold, and so was their land. It was this last, highly profitable enterprise that spurred the development of surveying techniques, already prepared by the military architects and artillerymen.

A year after the Dissolution a monk called Richard Benese produced a book on how to survey land using the simple tools of the time, a rod with a cord carrying knots at certain intervals, waxed and resined against wet weather. It was Benese who fixed the measure of an acre: 'an acre bothe of woodlande, also of fyldelande [heath] is always forty perches in length, and four perches in breadth, though an acre of woodlande be more in quantitie than an acre of fyldelande.' Benese's system was simple – the field was divided into squares to be measured, and the sections that did not fall into neat squares were ignored. It was a rough and ready method, and good enough for selling church land in a hurry. For the whole of the sixteenth century there was a considerable trade in monastic land, and since most of the purchasers wanted to know exactly what they had bought, surveyors found themselves in a seller's market. Everybody wanted to employ them, and for a few years a lot of indifferent measuring went on. Digges' theodelitus helped, and the quality of surveying gradually improved. Already Laurence Nowell had written, in 1563, to Secretary of State William Cecil proposing maps of the English counties because those already in existence were so bad. As the new process of copper engraving came from the Low Countries, the old estate-surveyors' written reports gave way to new maps. Christopher Saxton was the first surveyor to produce and publish accurate maps of England, in his beautiful atlas of the country published in 1579, followed in 1584 by a smaller map of England and Wales. For this, the first national atlas in Western Europe, Saxton became the only mapmaker to be granted a coat of arms and land by a reigning monarch. He was followed by John Norden, who was thirty-one when Saxton's

atlas appeared. His county maps, first produced in 1592, marked the principal roads for the first time. Norden and other surveyors were encouraged by Cecil, who was conscious of the defensive weaknesses of the realm, and who demanded a map every time he was called upon to make a decision relating in any way to land, harbour construction, drainage, or coastal fortifications.

On the Continent, too, the military needs of the war in Flanders had produced a further rush of fortifications in the adjacent French territory under Henri IV and his superintendent of fortifications, the duc de Sully. Sully had appointed engineers in charge of each province, and with them 'designers' whose job it was to map the local areas. By the time he fell from power in 1610, Sully had obtained the best possible maps of the high-risk provinces along the coast and the border with Flanders. Although John Norden had put roads on his maps of 1592, and Saxton's maps were republished much later with roads added, the first systematic illustration of roads on large-scale maps came in 1632 on the French post-road maps of Nicholas Sanson, of Abbeville in Normandy, and from the great school of cartography that he founded. In 1658 he published a collection of maps 'of all the parts of the world'. Further stimulus to the technique of surveying was

Saxton's map of the English county of Norfolk (1574), containing, as is claimed top left, 26 market towns, 625 villages, and all rivers 'truly described'. Saxton's coat of arms is displayed bottom left, and the scale, in miles, bottom right. Note the first crude attempts to show relief.

264

provided ten years later by the Fire of London. Much of the city had to be re-surveyed, since apart from those houses destroyed by fire many had been deliberately demolished in order to create fire-breaks. Furthermore, the increase in overseas trade at the time, especially with the colonies in America and with the Far East, demanded considerable financial backing, and mortgages were being raised on land in order to find the money. This too gave work to the surveyors.

The greatest map-maker of the seventeenth century was undoubtedly John Ogilby, appointed 'sworn viewer' of London after the fire. Unfortunately Ogilby's 'Britannia' atlas was never completed after his death in 1676, but by that time he had completed the road-map sections. These contained details of all main roads, and distances were for the first time measured in statute miles of 1760 yards. As a result, the distance between London and Berwick upon Tweed changed from 260 miles to 339. For the purpose of exact measurement Ogilby used a foot-wheel, with a counter to total the number of wheel revolutions between points. The year before Ogilby died, the French astronomer Jean Picard had begun work on

The first plate of Ogilby's road map from London to Berwick, on the Scottish border, giving statute mileage between London and the principal towns en route. Note the addition of a compass showing north on every section of road, and the use of buildings to give an idea of the size of towns and villages, as well as descriptions of local features to guide the traveller.

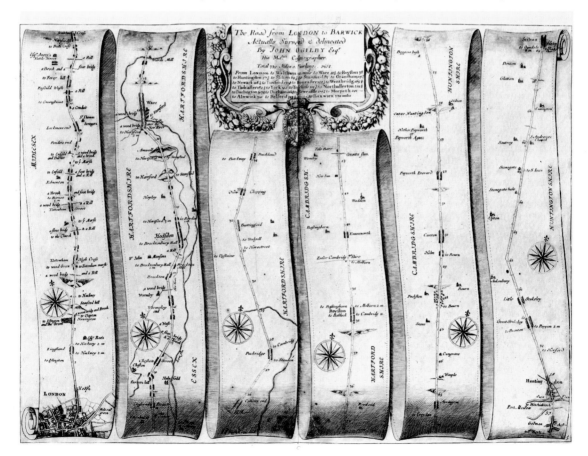

the shape of the Earth. (It was Picard whose observation of the mercury glow in his barometer had started off the chain of investigation which was to end in the invention of electricity generators and telecommunications.) He wanted to know whether the Earth was round, pear-shaped or egg-shaped, and in order to find out he was arranging expeditions to distant places like Lapland, where the meridian lines would be much closer to each other as they approached the poles than they were, for example, in France. Comparisons would then reveal whether these meridians described a sphere, an egg or a pear. This involvement on the part of French astronomy was to have repercussions later for the English map-makers.

Meanwhile detailed work continued on county surveys, which extended as far as the Lowlands of Scotland. This limitation was to prove acutely embarrassing when, in 1715, the Stuart family made their bid for the English throne. The 23-year-old James Edward Stuart claimed to be, by right, James VIII of Scotland and James III of England, and with French help and much support from the Highland Scots he caused his standard to be raised on the Braes of Mar in Aberdeenshire on 6 September of that year. The rebellion against the new German King George in London was easily crushed. It was badly organized and ineptly led, but it taught the English generals that war cannot be conducted easily where there are no maps – and no one had previously thought to survey Scotland. In spite of this realization nothing was done until 1724. In that year, following reports that the Highlanders had not been quite as pacified by the measures taken following the rebellion as might have been thought, the Member of Parliament for Bath, a Major-General George Wade, was sent to examine the situation.

According to Wade, things were bad: of the 22,000 clansmen fully 12,000 could be expected to take up arms against the English throne should another opportunity present itself. He suggested that garrisons should be stationed permanently in the area for the purpose of keeping an eye on the clans, and that a system of roads should be built to connect these garrisons, so that they could be moved quickly in aid of each other when the occasion demanded. Work on the roads began the following year, beginning with the stretch between Fort William and Fort Augustus. Wade's road-builders were soldiers taken from various companies, who were kept working by what, for the times, was an unusually democratic spirit in the ranks, as well as double pay and frequent celebratory ox-roasts. Although their instruments differed little from those of Digges' time, the road was a remarkable achievement. The section over Corryarrick Pass, which may be walked today, makes no fewer than eighteen traverses with hairpin bends, provided with underground culverts to drain away water in bad weather. Construction work was particularly thorough. The surface was a mixture of gravel and tar resting on a bed of stones and boulders. Either side of the road was a raised kerb, with a drainage ditch outside it. Where the road ran along the side of a hill, shallow drains were cut diagonally across its surface and floored in stone. Wade's roads were of exceptional quality, and would have been perfectly suited to the automobile had they not been virtually destroyed by the coaches and horses that were next to use them. At the time the saying went: 'If you had seen these roads before they were made, you would throw up your hands and bless General Wade.' Wade finished over 250 miles of roads in 1731, was promoted to full general, and went south in 1740.

Five years later he was back, chasing Bonnie Prince Charlie along the very roads he had built. Ironically it was Wade's road that Charlie used on his way south when he advanced as far as the town of Derby, less than 130 miles from London. His arrival threw the capital into uproar. The Bank of England would pay only in sixpences, and George II packed his bags ready to move back to the family home in Hanover. Charlie could not be persuaded to make his final move and take London. He withdrew to Scotland, and soon Wade and the Black Watch were once more on the offensive. The English populace cheered them on. Nightly at Drury Lane, the theatre audiences sang Wade's praises in a verse specially added to the National Anthem:

> God grant that Marshal Wade
> May by thy mighty aid
> Victory bring.
> May he sedition crush
> And like a torrent rush
> Rebellious Scots to crush
> God save the King.

In 1746 Charlie lost at the battle of Culloden, but what he had almost succeeded in doing left an indelible impression on a young post office draughtsman called William Roy, who became involved in the abortive mapping of Highland Scotland which the English were finally persuaded to attempt but which, even now, they were destined not to finish.

In 1747 Roy began at Fort Augustus, ready to map every inch of the Highlands, but by 1755 war with France had stopped the work. Roy was recalled to make military maps of areas of Kent and Sussex most vulnerable to invasion by the French. The war of 1755 lasted until 1763, and in the years immediately following it Roy – who was by now in the Army – moved gradually up the ranks. In 1783 he was promoted colonel and director of the Royal Engineers, and given a

major task. By this time the work begun by Picard in France had reached the stage where the French were preparing to plot the exact distance between the Greenwich and Paris Observatories so as to fix the lines of meridian that passed through them. Jean Cassini, now in charge of the project, wrote in 1783 to suggest the detailed triangulated mapping of south-east England, and Roy was given responsibility for the work. It was made possible by the development of Jesse Ramsden's giant theodolite, fitted with telescopic sights and capable of measuring with accuracy down to one minute of a degree, thanks to Ramsden's new method of marking the scale. In 1787 the English base was set up outside London on Hounslow Heath, and, after the most exact possible identification of where the spot was on the face of the Earth, work was begun. The French Revolution stopped it – but it was fear of possible invasion from the new Republic that moved the British, in 1791, to set up the Ordnance Survey, and begin to map the entire country using the latest surveying techniques.

Then, in 1824, the Survey went to Ireland. The reason for the move was a financial one. For years the British Government had complained at the inadequate amount of taxes it was receiving from Ireland. These taxes were used to pay for the upkeep of gaols, courts, roads and bridges, and for official salaries, and were assessed on the rateable value of property. However, different parts of Ireland used different methods of assessment, different measures, and different maps. Not one town boundary had yet been officially established. When the Lord Lieutenant of Ireland, Lord Wellesley, wrote to the Duke of Wellington to point out that Ireland did not have enough surveyors

The Ramsden theodolite has a 3-foot horizontal circle of brass, carrying four microscopes which permit it to be read to within one second of arc. The main telescope is made accurate horizontally with two spirit levels mounted alongside. The instrument was accurate to within 5 inches over a distance of 70 miles.

to carry out the work, the duke set up a school at the Woolwich Arsenal in London, where twenty cadets were to be trained for the job. One of those cadets was a young graduate of Edinburgh University called Thomas Drummond. At the age of sixteen, having studied maths, natural philosophy and chemistry, he left the university to become a Woolwich cadet. Two years later, in 1815, he entered the Royal Engineers. In 1819 he met a Colonel Thomas Colby, who interested him in the work of the Ordnance Survey. It was Colby who, in 1825, encountered a particular problem on the top of Divis Mountain, outside Belfast.

The Survey of Ireland had begun with a preliminary reconnaissance, and since the country was to be surveyed using triangulation it was necessary that high terrain be used in order to achieve long-distance point-to-point measurement. The longest side of the first triangle to be established ran between Divis, and Slieve Snaght in Donegal. Once the triangle they formed with a point in Scotland was known exactly, it would be possible to take bearings from both mountains to other points in the country and work out the distances from them geometrically. The problem was that it was autumn, and the weather was bad. Try as they might with flames and various kinds of light, little could be seen of Slieve Snaght at all through the murk and rain, let alone a pinpoint of light on its summit. Then Drummond produced something he had been working on for much of the previous year. It was a new method of producing light, and Drummond was packed off to Slieve Snaght to try it. In a letter to his mother he described his arrival there, on 27 October: '. . . for the first week our life was a struggle with the tempest – our tents blown down, our instruments narrowly escaping, and ourselves nearly exhausted. At length, by great exertions, we got two huts erected, one for the

Surveying on the grand scale, 1906. An intrepid band of surveyors with their theodolite on top of Monte Leone, on the Swiss–Italian border, during the preparations for the construction of the Simplon Tunnel under the Alps. This, the highest point in the area, served as the centre of triangulation operations.

seven men who are with me, and the other for me.' As soon as they were ready Drummond lit the new light. In the first stage of operation a jet of oxygen was produced through a nozzle, the tip of which was placed in alcohol. When the alcohol was ignited it burned fiercely in the oxygen-rich atmosphere produced by the jet of gas. The flame was then played on a small ball of lime, three-quarters of an inch in diameter. As the lime heated it began to give out a brilliant white light – Drummond claimed it was eighty-three times as bright as the best conventional lamps – and this light was reflected in a parabolic mirror placed behind the flame. The result was a narrow ray of intense white light. The apparatus was known as limelight, and it was an instant success. On 9 November a messenger came struggling up the mountain to tell Drummond to put it out; the observers had seen it clearly from the instant Drummond had lit up, and had taken the necessary readings. Divis Mountain and Slieve Snaght were $66\frac{1}{4}$ miles apart.

Drummond got his reward not long afterwards, when he demonstrated his new limelight before the senior scientists of the day. The 300-foot long Armoury in the Tower of London was chosen for the exhibition. First came the Argand gas burner, with its parabolic

Although the Drummond lime-light was used principally in lighthouses and theatres it was also tried briefly as street lighting. The Illustrated London News *of May 1860 shows the lights mounted experimentally on Westminster Bridge, alongside the much dimmer gaslights of the time.*

reflector – the type in common use at the time in British lighthouses. Then a light played through the new Fresnel lens, which made the Argand dim by comparison. And finally Drummond's light. In the words of Sir John Herschel, the astronomer: '. . . the lime being brought now to its full ignition and the screen suddenly removed, a glare shone forth, overpowering, and as it were annihilating both its predecessors . . . . A shout of triumph and of admiration burst from all present.' The emotion displayed may have owed a little to patriotism, since, after all, the best lights available till then were both French. With memories of the end of the Napoleonic wars still relatively fresh, it is perhaps not surprising that a nation so dependent on overseas trade and control of her sea routes should have jumped at the possibility of replacing the French Argand lights in British lighthouses with a home product. Be that as it may, the next demonstration was for the Masters of Trinity House, the authority responsible for lighthouses.

The results of this, held on 5 March 1830, were encouraging, and the apparatus was moved to a temporary lighthouse at Purfleet in the Thames estuary for trials. On 10 May the observers installed themselves on Trinity Wharf, Blackwall – ten miles away from Purfleet, but in direct line of sight of it. For this test, Drummond had changed his ignition mixture to hydrogen and oxygen, which gave off an even brighter flame. On the evening of 25 May the light went on, and cast shadows of the observers! The event 'elicited a shout of admiration from the whole party . . . The light was not only more vivid and conspicuous, but was peculiarly remarkable from its exquisite white-ness. Indeed there seems no great presumption in comparing its splendour to the sun . . .' In an air of great anticipation further tests were conducted on established lighthouses, including that on South Foreland. But the gas pipes leaked, supplies were inadequate and bags of gas burst. The keepers wrote logs of the trials: 'Thursday night, 27th. Limelight discontinued, in not getting sufficient gas. 2nd September. Bricklayer fixing two more retorts. A great escape from the gasholder.' And so it went on. In the end Drummond's light was turned down on the grounds of expense. He had to look elsewhere.

It was trouble with the new coal gas which gave him the opportunity he was waiting for. Since its introduction a few years before, gas had become very popular as an illuminant in theatres. Covent Garden had taken the lead in 1815. But the gas doubled the number of theatre fires in the decade that followed. The sheer complexity and unwieldiness of the supply systems made fires almost inevitable. In the Paris Opéra, for example, there were no less than 28 miles of tubing attached to a control panel, referred to as 'the organ', on which were 88 taps used to direct gas flow to 960 gas-jets, switched on and off according to detailed lighting plots from the show's producer. With so many joints and connections, leaks were plentiful. As the magazine *The Builder* remarked, in 1856: 'The fate of a theatre is to be burned. It seems simply a question of time.' The attraction of the brilliance of

Gaslight in use at a pantomime rehearsal, 1881. The portable gas burner shown here would be removed for the performance, when individual gas footlights would be used. Until the development of the gas mantle in 1893 the light produced by gas was relatively poor, and the use of such extra burners greatly increased the risk of fire in theatres.

limelight was too much for the theatre proprietors. By 1837 Clarkson Stanfield was using it to give extra effects on the Diorama of Continental Views in his pantomime. The new light won extravagant praise when it was used to spectacular effect in the 1851 Drury Lane production of *Azael*. Henry Irving was the only famous artist to be criticized for his use of the light. The reviews of his 1889 production of *Macbeth* sneered at 'a strong shaft of limelight obviously proceeding from nowhere'. Nonetheless, limelight became used extensively for creating realistic beams of moonlight or sunlight, as well as for isolating actors in a lime spotlight – for which effect it has passed into our language as a synonym for fame.

Meanwhile others were trying to solve Drummond's original problem of gas supply to the limelight so that costs would fall enough for it to be of practical use in lighthouses. The British authorities were still keen to improve the lights round the coastline, since as shipping increased so did losses. By mid-century these were averaging over 800 a year. In 1849 Floris Nollet, a Belgian Professor of Physics at the Ecole Militaire in Brussels, hit on the idea of generating hydrogen and oxygen by electrolysis. It was known that if a current of electricity were passed from a positive to a negative electrode in water, it would cause the molecules of the water to break apart into their constituents, hydrogen and oxygen, and these would be given off in the form of gas bubbles which could be collected. All that was

A Nollet-type generator in use in Montmartre during the siege of Paris (1870) to power an arc searchlight, which is being directed on a target observed by the soldier looking out of the window through binoculars. There are two more arc lamps on the table, on either side of an oil lamp. In the background can be seen the steam engine powering the generator.

The Holmes magneto-electric generator on show at the Great Exhibition in London, 1862. The current was picked up by the rotating discs above the main shaft at the centre, and conducted away down the wires attached to the disc supports.

needed was a very high current. Nollet knew that if a metal disc were rotated between the poles of a horseshoe magnet, electricity would be produced. (This had been shown by Faraday as early as 1831.) Nollet designed a generator composed of banks of magnets, between whose poles the discs would rotate. After his death in 1853 a company called the Société de l'Alliance was formed to exploit his patent. One of the engineers on the project was an Englishman called Frederick Holmes, who returned to England in 1856 convinced that the generator would serve an altogether different purpose from producing gas.

What he had in mind was supplying current directly to an arc light, which until then had suffered from the limitation imposed by its need for a great deal of power. The arc principle was a simple one: current was sent along rods of carbon, which were placed with their tips almost touching. At the tips the current jumped the gap, creating a spark which caused the carbon to become incandescent. In 1857 Holmes approached the Trinity House authorities with his new generator. It stood 9 feet high, and in its final version consisted of three vertical circles of twenty magnets. In the space between the circles two rings of discs, mounted on a central, horizontal shaft driven by a steam engine, rotated between the magnets ninety times a minute.

This arrangement produced the amount of current it takes to run a modern domestic electric fire, and weighed $5\frac{1}{2}$ tons! After several trials in lighthouses at South Foreland and Dungeness, the system was installed in the new Souter Point lighthouse on the Northumberland coast, in 1871. It was already obsolete. The year before, in Paris, a French engineer called Zénobie Gramme who had also worked on Nollet's ideas in Belgium, produced the first commercial version of an idea put forward some years earlier by a Dane, Sören Hjorth. The idea was that if a soft iron bar were induced with current from a coil of wire round it, it would become a magnet. If this magnet were then spun within the wire coil, it would induce current in the wire, which would in turn make the iron more magnetic; this would build the current in the wire, which would make the iron more magnetic, to the point where at a certain speed the iron would become saturated with magnetism. When this happened, the maximum amount of current was being produced, and would go on being generated as long as the machine was left working. Since the device produced electricity 'dynamically', it was called a dynamo-electric generator. This was later shortened to dynamo. Gramme's dynamo was the first to produce truly continuous current, and its arrival finally clinched the success of the arc light.

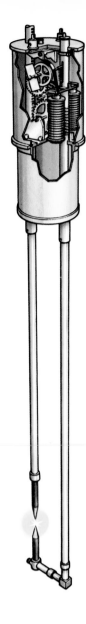

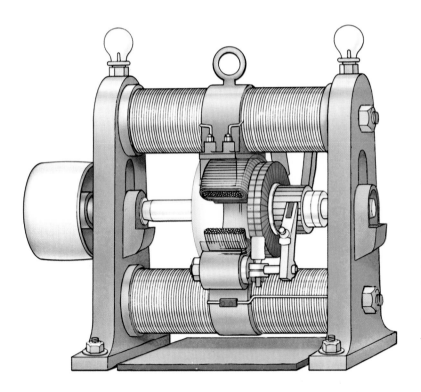

Left: The Gramme dynamo, produced in 1877 for use with arc lights. The armature (the wire coil) is in the centre, cut away to show its construction. This ring armature spins within the magnetic field produced by four electromagnets (above and below the armature). As the armature spins it cuts the magnetic field first at its north pole, then its south, producing positive– negative alternating current.

Left: The Crompton arc light. The bright glow occurs when an electric current is passed through the tips of the carbon rods, separated by about 1/8 of an inch. At this point the tips become incandescent and begin to burn away. The correct separation is maintained by two cylinder-shaped electromagnets.

Right: A special issue of the Illustrated London News *of 1879 celebrates the new era in a picture proclaiming the 'Triumph of steam and electricity'. The scenes on the left were meant to represent the world before science had changed it, although (bottom left) this somewhat belittled the enormous effect of gaslight at the beginning of the century.*

The process by which innovation and change comes about has already been seen to be influenced by such varying factors as climate, political expediency, investigation, accident and so on. What was to happen next is a fine example of how innovation sometimes occurs because at a certain point all the elements of a new development become available to one man, who fits them together like the pieces of a jigsaw. The arc light is one of the pieces. The second piece began to take shape in the mountains of southern central Europe, where gunpowder used for blasting was in increasingly short supply. Because one of its ingredients was charcoal it relied on dwindling sources of wood, the price of which had become unacceptably high. Besides, gunpowder tended to throw rocks high into the air but was not powerful enough to fragment them. In May 1846 Christian Schönbein, a Professor of Chemistry at Basle University, announced that he had found an alternative explosive. It was made by immersing cotton in a mixture of nitric and sulphuric acid, in a ratio of one to three, pressing the cotton to squeeze out the acid residue, and allowing it to stand for an hour. At this point the impregnated cotton was washed in water, pressed and dried. When lit, this 'gun cotton' would explode. Schönbein described its effectiveness: 'One pound of gun cotton is as effective as two to four pounds of black [gun] powder . . . cotton so treated does not leave any residue when exploded, and produces no smoke. The manufacture is not attended with the least danger'. This last remark interested John Hall (the man who earlier

The result of the Faversham explosion, which was heard seventeen miles away. By a curious irony, the factory was on a site which had been previously occupied by a gunpowder-making concern for more than a hundred years.

had been involved with Bryan Donkin in the production of the first canned food), and he came to an arrangement with Schönbein through which he could produce the cotton. In 1847 he leased a factory near Faversham, Kent, where explosives had been manufactured since the fifteenth century, and began production. By some accident – in all probability allowing the temperature of the processed cotton to rise too high – there was a gigantic explosion at the factory at 11 a.m. on 14 July 1847. An eyewitness report appeared in *The Times*: 'The roofs of all the buildings within about a quarter of a mile of the explosion are completely stripped of their tiles, and the walls are much shaken. Even in the town of Faversham, fully a mile distant from the scene of the disaster, windows were broken.' The factory was destroyed, and twenty-one people were killed – ten of them blown to pieces. The event stopped the production of gun cotton for the next twenty years.

Then the American firm of Phelan and Collander, makers of ivory billiard balls, ran into a severe shortage of their raw material, and offered a prize of $10,000 for the development of any adequate substitute. The offer excited the interest of a young printer living in Albany, New York, called John Hyatt. In 1870 he obtained a patent for the new material. It was made of gun cotton mixed with alcohol and camphor, heated to make it pliable, moulded, and allowed to harden. There is no record of whether or not Hyatt won the prize, but there were unconfirmed stories of odd events in Western saloons when the new billiard balls would strike each other with a noise like gunshots – a stimulating sound in a room filled with men carrying guns. Hyatt's billiard ball forms the second part of the jigsaw.

The third part began life in Vienna, in 1845, at the hands of an artillery instructor called Franz von Uchatius. By 1853 he had per-fected his technique for teaching ballistic theory to his students sufficiently to give a demonstration to the Vienna Academy of Science. He had taken the contemporary idea of stroboscopic light effects, and adapted it. His method was to paint on to a transparent glass disc twelve pictures, each showing different stages of the move-ment of a shell through the air. In front of each picture was a small lens, and the twelve lenses were angled so that they all focused at the same point on a screen or wall. Behind the pictures, which he kept stationary, he rotated a source of light – he used a Drummond limelight – which was shielded so as to light only one picture at a time. As he cranked the light round and round, so each picture was flashed on to the screen in sequence. The faster he cranked, the more the shell appeared to move. Living in Vienna at the time was a famous magician called Ludwig Dobler, who was so impressed with the possibilities of Uchatius' idea that he bought it on the spot for a hundred florins. Within a few years the 'dissolving pictures' were the rage of Europe.

In 1872 the Governor of California, Leland Stanford, who was a racing fanatic, had an argument with a friend of his called Frederick MacCrellish as to whether or not a horse at the gallop touched the ground with only one foot. Stanford claimed it did, MacCrellish that it did not. It happened that Stanford knew an immigrant English photographer, Eadweard Muybridge (who had changed his name from Edward Muggeridge). Muybridge was engaged in photographing Mexico and Central America for the Central Pacific Railway, in which Stanford had some interest. Stanford asked him if he could work out a way to settle the argument. The investigation took place on 15 June 1878 at the Governor's training course in Palo Alto, California, and newspaper reporters were invited to see fair play. Muybridge had set up twelve cameras along the side of the track, at 40-foot intervals because Stanford's horse was reckoned to gallop at 40 feet a second. The camera shutters, developed by John Isaacs, an engineer on the railway, were activated by a signal triggered as the horse broke through a series of twelve threads stretched across the track in its path. Opposite the cameras on the other side of the course stood a 15-foot high fence, marked off in vertical lines spaced at intervals of 21 inches and numbered consecutively so that the processed photographs could be assembled in the correct order. The photographs became world famous, patented by Muybridge as a series of postcards. They proved the Governor wrong. The following year Muybridge demonstrated his 'zoopraxiscope': it consisted of a central light source illuminating transparent glass discs, each of which was

Above: One of the zoopraxiscope discs which introduced the idea of serial motion to the world.

Left: A series of photographs taken by Muybridge of a galloping horse. The last three pictures in the top row show that all the horse's hooves are off the ground during the gallop.

illuminated through a slit cut into an opaque disc set behind the glass. On each glass disc was a painting taken from Muybridge's photographs. As the discs revolved, the photographic subject appeared to move. The zoopraxiscope is the third part of the jigsaw.

The fourth part is a direct descendant of the morse code receiver, which uses an electromagnet to attract or free a metal bar in response to electric pulses coming into the apparatus from a telegraph wire. As the bar engages and disengages it makes a tapping sound, coded in the manner called after the painter who claimed to have invented it. In 1877, at Menlo Park, New Jersey, a man who had been a telegraph operator earlier in his life produced a means of storing the messages and passing them on as desired, automatically. The stored message was recorded on a waxed paper disc, in which indentations were made by a stylus that moved up and down in response to the incoming signals, tracing a path of indentations as the disc revolved on a turntable, much like a modern record player. To replay the message simply involved turning the disc upside down – which had the effect of turning the indentations into small bumps. Another reader stylus was placed on these bumps, and as the needle tracked across the bumps it moved up and down, making and breaking contact with a switch that relayed a signal every time contact was made. The pattern of contacts was the same as the pattern of the indentations made by the incoming signal in the first place. The story goes that the man who developed this repeating telegraph – Thomas Edison – heard it going at high speed one day and fancied it gave off a varying, musical tone, and so realized that there was a connection between the vibrations of the stylus and a sound pattern. He hit upon the idea (not a new one) that he could reproduce sound as vibrations by the use of a thin membrane, so he placed a needle on the membrane that would bump up and down as the membrane vibrated, and mounted it in such a way as to score a bumpy path in tinfoil. The space available on which to score the track was increased by rolling the tinfoil round a cylinder set on a screw shaft, so that it advanced a little each time it was turned, and the needle moved over fresh tinfoil with each revolution. Once he had this bumpy path, which he produced by shouting 'Mary had a little lamb, its fleece was white as snow' at the membrane, he returned the cylinder to its original position, placed the needle in the groove, and turned the cylinder at the same speed he had used for the recording. The needle bumped along the groove, in the same pattern as it had previously made, causing the membrane to vibrate exactly as it had done in response to Edison's voice, and from the membrane came a voice reciting the rhyme. Edison called his development the phonograph. When the patent was issued, on 15 December 1877, it made him both rich and almost immediately famous throughout the world.

The repeating telegraph, developed by Edison because as an operator he had frequently been overwhelmed by message traffic.

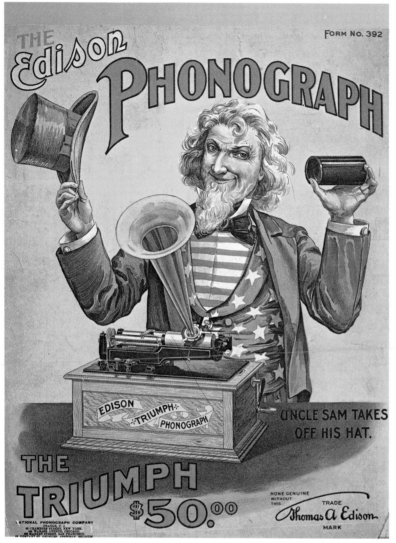

An early advertisement for Edison's phonograph which appealed to patriotism – it was accompanied by the caption 'Uncle Sam takes off his hat'. Nor was the choice of the name Triumph accidental. Edison marketed his 'inventions' as carefully as he developed them, and as assiduously as he cultivated the image (below) of the lonely genius, sleepless and exhausted from his labours.

The first part of our jigsaw, the arc light, was to undergo one more change at Edison's hands. He, like the men who had worked on the arc light, realized that carbon was capable of taking a great deal of heat while burning only very slowly. For fifty years people had been trying to solve the problem of the carbon burning away by placing it in a vacuum, but none of these vacua were efficient enough, and even the presence of minute traces of air caused the carbon to oxidize. In 1878 Edison began seriously to try to find a solution. He started by trying metal filaments, such as platinum. Either they oxidized, or were too expensive to consider. He returned to carbon, placing strips of paper in a kiln together with carbon and producing carbonized paper filament. They burned out too. Then a new pump arrived from

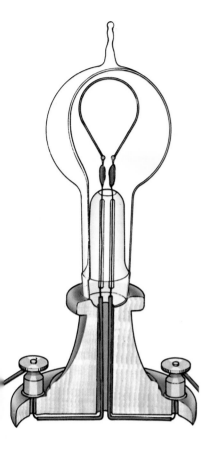

Germany. It had been made by a man called Hermann Sprengel for the investigation being carried out in Europe into the behaviour of electric currents in vacuum tubes. With the Sprengel pump, Edison had an almost perfect vacuum in his glass bulbs. After patiently testing 6000 different materials as filaments, he returned to carbon. In October 1879 a carbonized cardboard filament stayed lit for 170 hours, and Edison and his team stayed up night and day to watch it.

With the light bulb, Edison had the final piece of the jigsaw: the other three were the billiard ball, the zoopraxiscope, and his phonograph. The most essential element was, perhaps, the billiard ball. The Hyatt brothers had given their new material from which the ball shell was made the name of celluloid, and by 1888 it was already in use as film, in stills photography. The year before an obscure clergyman, Hannibal Goodwin, had filed a patent for producing celluloid film, but was repeatedly taken to court by one George Eastman, and finally died impoverished. Eastman's company produced celluloid film in 1889. As for the zoopraxiscope, Muybridge and Edison had known each other since 1886, when the former had visited Edison at his laboratory after giving a lecture on his invention. It was almost inevitable that Edison should one day put together the light, the film and the concept of moving pictures. He did so in 1893 in his kinetoscope, with special strip film produced for him by Eastman, and a ratchet system for pulling the film through the holder by means of small holes in each side of the strip. The kinetoscope was a great success, but the concept of projection did not at once occur to Edison; it was patented by two Frenchmen, the Lumière brothers, in 1895.

Above: The 1881 Edison carbon filament electric lamp. Edison's real achievement lay in designing a power supply system that for the first time enabled lamps powered by the same generator to be switched on and off individually.

Right: His kinetoscope, the forerunner of the cinema. Edison designed a cog and claw system to pull the film forward past the lamp one frame at a time, rapidly enough for the eye to see serial motion.

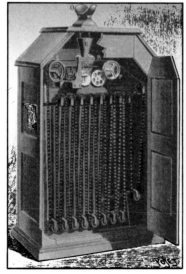

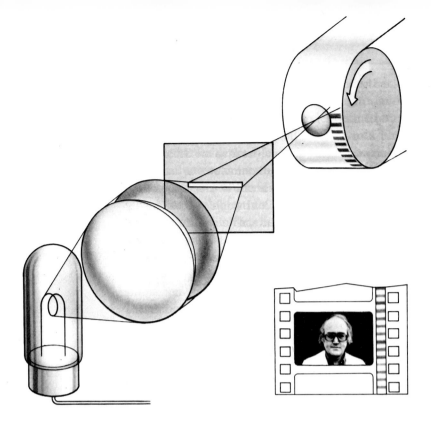

The new moving pictures proved so extraordinarily successful that attempts to turn them into 'talkies' were made almost immediately. Although the first smash-hit sound movie – *The Jazz Singer*, starring Al Jolson, released in 1927 – used a synchronized phonograph record as its sound source, the system which was to take over shortly afterwards involved the use of light. The idea may well have been inspired by Edison's 'vibration' reproduction in the phonograph. It was made possible by the use of a material whose properties had been known since 1873. The material was called selenium, and it reacted to the presence or absence of light by giving off a greater or smaller number of electrons. In 1923 a system was developed in Denmark which made possible the recording of sound directly on to film.

The idea that selenium and other metals were sensitive to light led to the development of one of the modern world's most ubiquitous and influential inventions. The major breakthrough in its development was made by a Russian émigré to the United States called Vladimir Zworykin. In 1928 he produced his iconoscope, which consisted of a glass container from which all air had been evacuated. Inside the container was a thin mica sheet; on the back of the sheet was a thin coating of metal, and on the front thousands of tiny particles of silver which had been treated with caesium vapour and oxygen, so that each particle had a coating of oxides of silver and caesium. These particles were arranged in rows, and were sensitive to light. When light from an image was focused on to the screen, each particle gave off electrons, just as selenium did. When this happened, each particle became posi-

Sound on film. The signal from the microphone causes the lamp to flicker. The lamplight is focused through a slit so that a rectangular area of negative is exposed, to be printed later as a more or less opaque positive. The opacity relates to the amount of noise originally fed in from the microphone.

tively charged, the strength of the charge being related to the amount of brightness affecting the particle and the length of time the particle was illuminated. In a separate part of the vacuum container was an electron gun, pointing at the screen of particles. From this gun a stream of electrons was fired, and magnetically directed so as to pass across the face of the screen row by row, hitting each particle in turn as it passed along the row. When the first row of particles had been scanned the electron beam would return to begin scanning the second, then the third, and so on. As the beam passed over a particle, that particle's charge would suddenly alter by an amount relative to the brightness of the part of the image which had illuminated it. This difference in potential was picked up by the silvered backplate as a voltage, which was then transmitted to a separate device. Each tiny signal, then, was an electrical translation of part of the original picture whose light had fallen on to the particle. As the signals continued to be released by the particles being scanned, they were used to alter the strength of another beam of electrons being fired in the same back-and-forward scanning motion at a phosphorescent screen, in order to reproduce, in a series of brighter and darker spots in the phosphorescence, the image which the screen of particles had 'seen'. In essence, the iconoscope was the forerunner of the television camera, and made possible electronic transmission of pictures on to the cathode ray tube that was a television screen.

When Edison died he had over a thousand patents filed in his name, thanks to the work of the men in his laboratory. Each man was a specialist in his field, serving the needs of Edison's fertile imagination as well as his acute understanding of the seller's market in which he lived. Edison never developed an idea unless he knew in advance that it would be profitable. He laid down six rules for invention: define the need for innovation; set yourself a clear goal and stick to it; analyse the major stages through which the invention must pass before it is complete, and follow them; make available at all times data on the progress of the work; ensure that each member of the team has a clearly defined area of activity; record everything for later examination. He also made full use of market analysis before embarking on any major invention. If there was no market, he did not start. He advertized heavily, promoting both the invention and his own image as the sleepless and brilliant inventor. Edison's laboratory was therefore the world's first true industrial research laboratory, the forerunner of the giant organizations of today such as Bell Labs, or General Electric, or Westinghouse. It is Edison we must thank for the present-day mania for change. He said: 'I can never pick a thing up without wishing to improve it.' In time, 'improve' became 'replace', as the attractions of built-in obsolescence became more obvious to the salesman.

Edison, it may be said, invented the business of invention, and so introduced the idea of change for the sake of change. His approach to innovation and marketing has bestowed on modern industry the problem of continued growth. The system we have inherited, immensely complex and interdependent, performs a constant balancing act, tipped in one direction and then another by factors that alter daily: a strike here, a new discovery there, a political change, an accident, a variation in climate, a war, a rumour – all those factors that have affected the course of change throughout history.

We have seen that each one of the modern man-made objects that alters our world, and in so doing changes our lives, moulds our behaviour, conditions our thoughts, has come to us at the end of a long and often haphazard series of events. The present is a legacy of the past, but it is a legacy that was bequeathed without full knowledge of what the gift would mean. At no time in the history of the development of the millions of artefacts with which we live did any of the people involved in that development understand what effect their work would have. Is today, then, merely the end-product of a vast and complicated series of accidental connections?

The iconoscope, forerunner of the modern television camera, seen on the left of the illustration. It incorporates within a sealed, evacuated glass vessel both the light-sensitive metal plate facing the subject, and the cathode ray 'scanner'. As is shown, the scanner is returning down the plate for the second time, so the newer, upper part of the scan shows up strongly on the right compared with the fading image of the lower part, remaining from the previous scan.

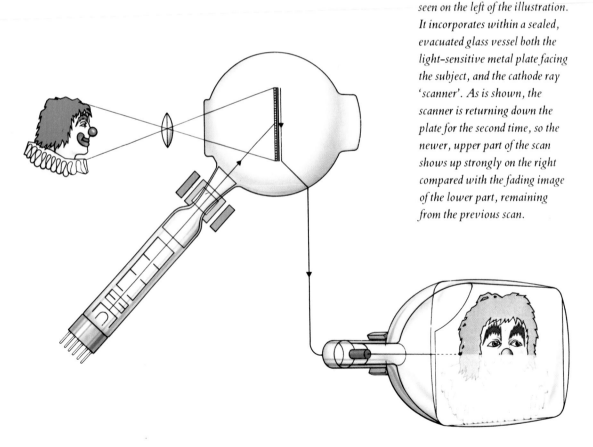

10
Inventing the Future

Why should we look to the past in order to prepare for the future? Because there is nowhere else to look. The real question is whether the past contains clues to the future. Either history is a series of individual and unrepeated acts which bear no relation to anything other than their immediate and unique temporal environment, or it is a series of events triggered by recurring factors which manifest themselves as a product of human behaviour at all times. If the latter is the case, it may be that the past illustrates a number of cause and effect sequences which may take place again, given similar circumstances. If it is not, then, as Henry Ford put it, 'history is bunk', and there is no profit to be had from its study, or from anything not immediately and only concerned with the unchanging laws of nature.

Clearly, a preference for the cause and effect argument governs the approach to history expressed in this book. The process of innovation is shown to be influenced by several factors recurring at different times and places; although these may not be repeated exactly each time, the observer becomes aware that they may recur in his own future, and is therefore more able to recognize them should they do so. The structural device used here is to examine an event in the past which bears similarity to one in the present in order to see where such an event led. Thus we return from the modern ballistic missile to the development of cannon-balls, from the telephone to medieval church postal services, from the atomic bomb to the stirrup, and so on. The purpose of this approach is to attempt to question the adequacy of the standard modern schoolbook treatment, in which history is represented in terms of heroes, themes or periods.

In the heroic treatment, historical change is shown to have been generated by the genius of individuals, conveniently labelled 'inventors'. In such a treatment, Edison invented the electric light, Bell the telephone, Gutenberg the printing press, Watt the steam engine, and so on. But no individual is responsible for producing an invention *ex nihilo*. The elevation of the single inventor to the position of sole creator at best exaggerates his influence over events, and at worst denies the involvement of those humbler members of society without whose work his task might have been impossible.

The thematic approach attempts to divide the past into subjects such as Transport, Communications, Sail, Steam, Warfare, Metallurgy and others, but this implies a degree of foreknowledge where none exists. Thus Bouchon's use of perforated paper in 1725 to automate the Lyons silk looms had nothing to do with the development of calculation or data transmission, and yet it was an integral part of the development of the computer. The Venturi principle, basic to the structure and operation of the jet engine or the carburettor, was originally produced in an attempt to measure the flow of water through pumps. Gutenberg's movable typeface belonged as much to metallurgy or textiles as it did to the development of literacy.

In the periodic treatment the past is seen as a series of sub-units bounded by specific dates such as the beginning of a new royal dynasty, the fall of an imperial capital city, the arrival of a new mode of transport, and these sub-units are conveniently labelled the Dark Ages, Middle Ages, Renaissance, Age of Enlightenment, and so forth. As this book has shown, such a view of the past is over-simplified, for to give any period a specific label is to ignore the overlapping nature of the passage of events. Elements of efficient Roman administration techniques continued to operate throughout the so-called Dark Ages. The fall of Constantinople meant little or nothing to the vast majority of the European population, if indeed they knew of it at all. There was no sudden and radical alteration in English life when the Tudors gave way to the Stuarts.

These approaches to the study of history tend to leave the layman with a linear view of the way change occurs, and this in turn affects the way he sees the future. Most people, if asked how the telephone is likely to develop during their lifetime, will consider merely the ways in which the instrument itself may change. If such changes include a reduction in size and cost and an increase in operating capability, it is easy to assume that the user will be encouraged to communicate more frequently than he does at present. But the major influence of the telephone on his life might come from an interaction between communications technology and other factors which have nothing to do with technology.

Consider, for instance, a point in the future at which depletion of energy resources makes Draconian governmental action necessary in order to enforce severe energy rationing. In such a situation the government might decide to tap telephones on a random national basis as part of a campaign to discourage profiteering and, in a more subtle manner, dissent. In these circumstances the telephone would become an instrument which would act as a brake rather than a stimulus to communication. This is typical of the way things happen. The triggering factor is more often than not operating in an area entirely unconnected with the situation which is about to undergo change. A linear view of the past would, for instance, place the arrival of the chimney in a sequence of developments relating to change in domestic living. Yet the alteration of life-style brought about by the chimney included year-round administration and increased intellectual activity, which in turn contributed to a general increase in the economic welfare of the community to a point where the increase in the construction of houses brought about a shortage of wood. The consequent need for alternative sources of energy spurred the development of a furnace which would operate efficiently on coal, and this led to the production of molten iron in large quantities, permitting the casting of the cylinders which were used in the early steam engines. Their use of air pressure led first to the investigation of gases and then petroleum as a fuel for the modern automobile engine, without which, in turn, powered flight would have been impossible.

Within this apparently haphazard structure of events we have seen that there are certain recurring factors at work in the process of change. The first is what one would expect: that an innovation occurs as the result of deliberate attempts to develop it. Napoleon presented the nation with clearly defined goals when he established the Society for the Encouragement of National Industry. One of those goals was the development of a means of preserving food, and it was reached by Appert's bottling process. When Edison began work on the development of the incandescent light bulb, he did so in response to the inadequacy of the arc light. All the means were available: a vacuum pump to evacuate the bulb, electric current, the filament concept which the arc light itself used, the use of carbon for the filament. The idea of serial motion in Muybridge's early photographs of the trotting horse led, through Edison's friendship with Muybridge, to the deliberate development of the kinetoscope as a money-making proposition. Von Linde perfected the domestic refrigeration system in answer to a specific request from the Munich beer brewers for a way of making and keeping beer in their cellars all year round.

A second factor which recurs frequently is that the attempt to find one thing leads to the discovery of another. William Perkin was in search of an artificial form of quinine, using some of the molecular combinations available in coal tar, when the black sludge with which one of his experiments ended turned out to be the first artificial aniline dye. Oersted's attempt to illustrate that a compass needle was not affected by electric current showed that in fact it was, and the electromagnet was the result of that surprise discovery. Henri Moissan, attempting to make artificial diamonds by subjecting common carbon to very high temperatures, failed to do so, but trying his luck with other materials at hand he produced calcium carbide, the basis for acetylene and fertilizer.

Another factor is one in which unrelated developments have a decisive effect on the main event. The existence of a pegged cylinder as a control mechanism for automated organs gave Bouchon the idea of using perforated paper for use in the silk loom. The medieval textile revolution, which was based on the use of the spinning wheel in conjunction with the horizontal loom, lowered the price of linen to the point where enough of it became available in rag form to revolutionize the production of paper. C.T.R. Wilson's cloud chamber gave the physicists the tool they needed to work on the splitting of the atom.

Motives such as war and religion may also act as major stimulants to innovation. The use of the cannon in the fourteenth and fifteenth centuries led to defensive architectural developments which made use of astronomical instruments that became the basic tools of map-making. The introduction of the stirrup, and through it, the medieval armoured shock-troop, helped to change the social and economic structure of Europe. The need to pray at predetermined times during the night and to know when feast days would occur aroused interest in Arab knowledge of astronomy. The water-powered alarm clock and the verge and foliot were the direct result of this interest.

Accident and unforeseen circumstances play a leading role in innovation. It was only when the bottom dropped out of the acetylene gas market that attempts were made to find a use for the vast amounts of calcium carbide in Europe and America: cheap fertilizer was the result. When the Earl of Dundonald's coal distillation kiln exploded and the vapours ignited, investigation into the gases resulted in the production of coal gas. The sudden arrival in Europe of the compass needle from China led to work on the phenomenon of magnetic attraction, and this in turn led to the discovery of electricity. A similarly unexpected Chinese invention, gunpowder, stimulated mining for metals to make cannon, and the money to pay for them. The flooding of these mines and the subsequent failure of the pumps brought about the development of the barometer.

Physical and climatic conditions play their part. As the European communities recovered after the withdrawal of the Roman legions and the centuries of invasion and war that had followed, reclamation of the land depended for its success on the development of a plough that would clear the forests and till some of the toughest bottom land in the world. These conditions helped to structure the mouldboard and coulter, implements that formed the basis for the radically new plough design that emerged in Europe in the ninth century, and thus helped to move the centre of economic power north of the Alps. The change in the weather which struck northern Europe like a sledgehammer in the twelfth and thirteenth centuries provided urgent need for more efficient heating. The chimney answered that need, and in doing so had the most profound effect on the economic and cultural life of the continent. In the early nineteenth century the prevalence of malaria in Florida, spread by the mosquitoes breeding in the swamps surrounding the town of Apalachicola, spurred John Gorrie to develop the ice-making machine and air-conditioning system in an attempt to cure his patients, in the mistaken belief that the disease was related directly to summer temperatures and miasma rising from rotting vegetation.

Two points arise from this way of looking at the process of change and innovation. One is that, as we have seen, no inventor works alone. The myth of the lonely genius, filled with vision and driven to exhaustion by his dream, may have been deliberately fostered by Edison, but even he did not invent without help from his colleagues and predecessors. The automobile, for example, was assembled from parts which included Volta's electric pistol, using the electric spark to ignite gases. Its basic piston and cylinder drive was Newcomen's. The carburettor owed its operation to Venturi's jet principle and its scent spray derivative. Its gears were descendants of the waterwheel. The elevation of the lonely inventor to a position of ivory-tower isolation does more than deny such debts; it makes more difficult the bridging of the gap between the technologist and the man in the street.

The second point is that the ease with which information can be spread is critical to the rate at which change occurs. The inventive output of Western technology can be said to have occurred in three major surges. The first – the Medieval Industrial Revolution – came after the establishment of safe lines of communication between the communities of Europe as order was re-established in the wake of the invasions of the tenth century. The second occurred in the seventeenth century when the scientific community began to make use of printing to exchange ideas on a major scale. The third followed the nineteenth-century development of telecommunications.

It was with the second of these stages that the age of specialization began, when scientists began to talk to each other in language that only their fellows could understand. The more the knowledge in a certain field increased, the more esoteric became the language. The reason for this is simply that ordinary everyday language has proved incapable of encompassing scientific subject matter. As the amount of knowledge in each field increased, the percentage of language shared only by others in the same field also grew. Today, ordinary persons are often prevented from sharing in scientific and technological discussions not by mental inadequacy, but because they lack certain key words and an understanding of their meaning. Has the rate of change become so high that it is impossible for the layperson to do more than keep up to date with the *shape* of things all around? It has been said that if you understand something today it must by definition be obsolete. And yet the rate at which change now occurs is an integral part of the way our society functions. The avalanche of ephemera that arrives in our homes every day has to continue to flow if the economy is to operate to the general advantage. Our industries are geared to high turnover, planned obsolescence, novelty. The basic components of the modern automobile have not changed in a generation, but minor modifications such as fuel-injection systems, suspension, lighting, seat cover material, body styling, have done so. Without these modifications the consumer would have no desire to change a machine which can operate efficiently, if well looked after, for decades or more. Thus change is good for the economy because it keeps the money going round and workers employed. The interdependence of such a society, however, renders it vulnerable to component failure. A small factory making one part of an automobile can bring the entire industry to a halt if its workers go on strike. The New York blackout of 1965 occurred because of the action of a single relay at Niagara Falls.

The production-line style of manufacture has increasingly affected our way of life, in some cases creating problems of alienation and dissatisfaction among wide sectors of the work force. But such difficulties go hand in hand with a higher standard of living engendered by these very production processes. As the pace quickens, and the diffusion of innovative ideas in the technological community is made easier by technological advance itself, the rate of change accelerates. At an obvious level this increases the avalanche of material goods and services provided, and makes life more comfortable – if also more complicated. But the amount of innovation increases also at an 'invisible' level – that at which a high degree of specialist knowledge is necessary to understand what is happening. Unfortunately, it is at this level that many of the advances most critically important to our future occur: developments in the field of genetic engineering, radioactive fuels, drugs, urban planning, and so on. It is in these areas of innovation that the average citizen feels disfranchized. He or she knows that they are happening, but does not know enough to be involved in decisions relating to them. Indeed, in some cases, people are actively discouraged from being so.

Thus the intelligent layperson realizes that he or she is surrounded by man-made objects – the products of innovation – that constantly serve as a reminder of his or her ignorance. It may be this contradiction, coupled with the possibility of resolving it provided by technology itself, which has led to a growing desire in the community to assess the likely future effects of the present high rate of change.

Attitudes have tended to fall into four groups. The first maintains that we should give up the present system oriented towards high technology and return to an 'intermediate' technology, making use of resources which cannot be depleted, such as wind, wave and solar power – in effect, that we should return to the land. The attraction of this idea is that it might draw society closer together by reducing the present gulf between the technologist and the citizen, thus encouraging participation on the part of the electorate in decision-making which would relate to simpler, more fundamental matters – a theory which has obvious appeal for the developing world, too. The question is whether such an alternative is possible for other than the simplest communities. Are we not in our advanced society already too dependent on technological life-support systems to make the switch?

The second attitude held is that we should assess scientific and technological research strictly according to its worth for society, and curtail all other forms of research. This presents more difficult problems. In selecting which areas of research to encourage and which to curtail, to what extent are we depriving ourselves of the benefits of serendipity, which, as this book shows, is at the heart of the process of change? Without Apollo, would we have had laptops? Without Moissan's search for artificial diamonds, would cyanamide fertilizer have been discovered? Without the atomic bomb, would fusion be feasible?

The third alternative is that we should allow technology to continue to solve the problems as it always has done. The result of this would be a continued rise in the standard of living, fuelled by cheaper consumer goods produced from cheaper power sources. Even if it worked, this option would still leave the community with two major concerns. The first is the effect on the environment of the vast amounts of virtually free energy created by the fusion process. While undoubtedly attractive in its stimulating effect on manufacture, the use of energy on such a scale might create a major planet-wide problem of heat. The waste heat created by the prolific use of this energy would have to go somewhere, and if it were not to have profound and long-term effects on the ecosphere, rationing measures might have to be taken. The next consideration is the inevitable frustration of the citizen should such rationing be enforced. This would surely increase, with potentially disruptive consequences.

The final alternative is that research and development should be directed towards producing more durable goods and less planned obsolescence. We should seek continued economic health in the markets as yet unsaturated by consumer products – in other words, 'spread the wealth'. This offers a utopian concept of the future. Sharing the present level of advance on a world-wide scale would present the manufacturing industries with decades of opportunity. The gap between the haves and have-nots would narrow, eventually to disappear, and with it would go the divisions that endanger the survival of man on the planet. Scientific and technological talent would be diverted to serve the greater ends of education in bringing the community together on a more equal material and intellectual footing. This vision is marred by the major dilemma it would present at its inception. How would it be possible to convince the haves that they had enough, and what would be the reaction of the have-nots if the richer communities would not accept a lower standard of living? The 1973 increase in oil prices demonstrated the power of countries which are less well developed economically but rich in raw materials to cause universal recession in a matter of months.

Whichever of these alternatives is chosen, the key to success will be the use we make of what is undoubtedly the vital commodity of the future: information. It seems inevitable that, unless changes are made in the way information is disseminated, we will soon become a society consisting of two classes: the informed élite, and the rest. The danger inherent in such a development is obvious.

In the meantime, we appear to be at another of the major crossroads in history. We are increasingly aware of the need to assess our use of technology and its impact on us, and indeed it is technology which has given us the tools with which to make such an assessment. But the average person is also aware of being inadequately prepared to make that assessment. Now that computer systems are within the price range of most organizations, and indeed of many individuals, an avalanche of data is about to be released on the man in the street. But what use are data if they cannot be understood?

In the last twenty years television has brought a wide spectrum of affairs into our living-rooms. Our emotional reaction to many of them – such as the problem of where to site atomic power stations, or the dilemma of genetic engineering, or the question of abortion – reveals the paradoxical situation in which we find ourselves. The very tools which might be used to foster understanding and reason, as opposed to emotional reflex, are themselves forced to operate at a level which only enhances the paradox. The high rate of change to which we have become accustomed affects the manner in which information is presented: when the viewer is deemed to be bored after only a few minutes of air time, or the reader after a few paragraphs, content is sacrificed for stimulus, and the problem is reinforced. The fundamental task of technology is to find a means to end this vicious circle, and to bring us all to a fuller comprehension of the technological system which governs and supports our lives. It is a difficult task, because it will involve surmounting barriers that have taken centuries to construct. During that time we have carried with us, and cherished, beliefs that are pre-technological in nature. These faiths place art and philosophy at the centre of man's existence, and science and technology on the periphery. According to this view, the former lead and the latter follow.

Yet, as this book has shown, the reverse is true. Without instruments, how could the Copernican revolution have taken place? Why are we taught that we gain insight and the experience of beauty only through art, when this is but a limited and second-hand representation of the infinitely deeper experience to be gained by direct observation of the world around us? For such observation to become significant it must be made in the light of knowledge. The sense of wonder and excitement to be derived from watching the way an insect's wing functions, or an amoeba divides, or a foetus is formed comes in its greatest intensity only to those who have been given the opportunity to find out *how* these things happen.

Science and technology have immeasurably enriched our material lives. If we are to realize the immense potential of a society living in harmony with the systems and artefacts which it has created, we must learn – and learn soon – to use science and technology to enrich our intellectual lives.

Further Reading

GENERAL

Bishop, Morris, *Penguin Book of the Middle Ages*
(Penguin: Harmondsworth, 1971).
Braudel, Fernand, *The Mediterranean and the
Mediterranean World,* 2 vols (Collins, 1949).
Cipolla, Carlo, *Before the Industrial Revolution*
(Methuen, 1976).
Crombie, A. C., *Scientific Change* (Heinemann, 1963).
Derry, T. K. and Williams, T. I., *A Short History of
Technology* (Oxford University Press, 1960).
Duby, Georges, *Rural Economy and Country Life in the
Mediaeval West* (Edward Arnold, 1968).
Hale, J. R., *Renaissance Europe* (Collins, 1971).
Harvey, John, *Medieval Craftsmen* (Batsford, 1975).
Krantzberg, M., *Technology in Western Civilisation*
(Oxford University Press, 1967).
Landes, D. S., *The Unbound Prometheus* (Cambridge
University Press, 1969).
Leff, Gordon, *Medieval Thought* (Penguin:
Harmondsworth, 1958).
Lopez, R. S., *The Birth of Europe* (Phoenix House,
1967).
Mathias, Peter, *Science and Society* (Cambridge
University Press, 1972).
Morgan, Bryan, *Men and Discoveries in Electricity*
(John Murray, 1952).
Pacey, Arnold, *The Maze of Ingenuity* (Allen Lane,
1974).
Parsons, W. B., *Engineers and Engineering in the
Renaissance* (MIT Press: Cambridge, Mass., 1939).
Pirenne, Henri, *Medieval Cities* (Doubleday: New
York, 1956).
Singer, C. *et al.* (eds), *A History of Technology* (Oxford
University Press, 1954).
Singer, C. and Underwood, E. A., *A Short History of
Medicine* (Clarendon Press: Oxford, 1962).
Street, A. and Alexander, W., *Metals in the Service of
Man* (Penguin: Harmondsworth, 1962).

CHAPTER 1

Aldred, Cyril, *Egypt to the End of the Old Kingdom*
(Thames and Hudson, 1965).
Breasted, J. H., *A History of Egypt* (Hodder and
Stoughton, 1946).
Lucas, A., *Ancient Egyptian Materials and Industries*
(Edward Arnold, 1926).
Moret, A., *The Nile and Egyptian Civilisation* (Kegan
Paul, 1927).

CHAPTER 2

Crossland, R. A. and Birchall, Ann (eds), *Bronze Age
Migrations in the Aegean* (Duckworth, 1973).
Johnson, F. R., *Astronomical Thought in Renaissance
England* (Octagon: New York, 1968).
Johnson, Paul, *Elizabeth: A Study in Power and Intellect*
(Weidenfeld and Nicolson, 1974).
Machlaurin, W. R., *Invention and Innovation in Radio*
(Macmillan: New York, 1949).
Mann, Martin, *Revolution in Electricity* (John Murray,
1962).
Neugebauer, O., *The Exact Sciences in Antiquity*
(Princeton University Press, 1952).
Parry, J. H. (ed.), *The European Reconnaissance*
(Macmillan, 1968).
Parsons, E. A., *The Alexandrian Library* (Cleaver Hume
Press, 1952).
Roller, Duane H., *The De Magnete of William Gilbert*
(Menno Hertzburger: Amsterdam, 1959).
Taylor, E. G. R., *The Haven Finding Art* (Hollis and
Carter, 1956).
Yass, Marion, *Hiroshima* (Wayland Publishers: Hove,
Sussex, 1976).

*Place of publication is London unless otherwise stated
(except in the case of university presses).*

CHAPTER 3

Barber, Richard, *The Knight and Chivalry* (Longman, 1970).

Bath, B. H. Slicher van, *The Agrarian History of Western Europe* (Edward Arnold, 1963).

Blair, Claude, *European Armour* (Batsford, 1958).

Bruce, R. V., *Bell: Alexander Graham Bell and the Conquest of Solitude* (Gollancz, 1973).

Butler, Denis, *1066: The Story of a Year* (Bland, 1966).

Cipolla, Carlo, *Guns and Sails in the Early Phase of European Expansion* (Collins, 1965).

Mesnard, Jean, *Pascal, His Life and Works* (Harvill Press, 1952).

Middleton, W. E. Knowles, *The History of the Barometer* (Johns Hopkins: Baltimore, 1964).

Tannahill, Reay, *Food in History* (Eyre Methuen, 1973).

CHAPTER 4

Febvre, Lucien and Martin, Henri, *The Coming of the Book* (New Left Books, 1976).

Gimpel, Jean, *The Medieval Machine* (Gollancz, 1977).

Laver, James, *Taste and Fashion* (Harrap, 1937).

Pitkin, Thomas M., *Keepers of the Gate* (New York University Press, 1975).

Przywara, Eric, *An Augustine Synthesis* (Sheed and Ward, 1945).

Rouse, Hunter and Ince, Simon, *History of Hydraulics* (Dover Publications, 1957).

Sumner, William L., *The Organ* (Macdonald, 1973).

Wallace-Hadrill, J. H., *The Long-Haired Kings* (Methuen, 1962).

Walton, Perry, *The Story of Textiles* (Tudor Publishing Company: New York, 1925).

White, L., Jr., *Machina Ex Deo* (MIT Press: Cambridge, Mass., 1968).

Ziegler, Philip, *The Black Death* (Collins, 1969).

CHAPTER 5

Ashton, T. S., *Iron and Steel in the Industrial Revolution* (Manchester University Press, 1963).

Cipolla, Carlo, *Clocks and Culture 1300–1700* (Collins, 1967).

Copley, F. Barkley, *Frederick W. Taylor*, Vols I, II (A. M. Kelly: New York, 1923).

Daumas, M., *Scientific Instruments of the Seventeenth and Eighteenth Centuries* (Batsford, 1972).

Grane, Leif, *Peter Abelard. Philosophy and Christianity in the Middle Ages* (Allen and Unwin, 1970).

Haskins, C. H., *The Renaissance of the Twelfth Century* (Harvard University Press, 1973).

O'Leary, De Lacy, *How Greek Science Passed to the Arabs* (Routledge and Kegan Paul, 1949).

Pannekoek, A., *A History of Astronomy* (Allen and Unwin, 1961).

Rolt, L. T. C., *Tools for the Job* (Batsford, 1965).

Rosenberg, N., *The American System of Manufacturers, 1854–5* (Edinburgh University Press, 1969).

Smith, Merritt Roe, *Harpers Ferry Armory and the New Technology* (Cornell University Press, 1977).

Woodbury, R. S., *History of the Lathe* (MIT Press: Cambridge, Mass., 1961).

CHAPTER 6

Cardwell, D. S. L., *Steam Power in the Eighteenth Century* (Sheed and Ward, 1963).

Day, Joan, *Bristol Brass: History of the Industry* (David and Charles: Newton Abbot, 1973).

Dibner, Bern, *Alessandro Volta and the Electric Battery* (Franklin Watts: New York, 1964).

Donald, M. B., *Elizabethan Monopolies* (Oliver and Boyd, 1961).

Gunston, Bill, *The Jet Age* (Arthur Barker, 1971).

Nixon, St John C., *Daimler 1896–1946* (Foulis and Co.: Sparkford, Somerset, 1947).

Quennell, M. and C. H. B., *Everyday Life in Anglo-Saxon, Viking and Norman Times* (Batsford, 1955).

Raistrick, A., *Quakers in Science and Industry* (David and Charles: Newton Abbot, 1950).

Rolt, L. T. C., *Thomas Newcomen: Prehistory of the Steam Engine* (David and Charles: Newton Abbot, 1963).

Sanderson, Ivan T., *Follow the Whale* (Cassell, 1958).

Savage, George, *Glass* (Weidenfeld and Nicolson, 1965).

Wood, Margaret, *The English Medieval House* (Dent, 1965).

CHAPTER 7

Albion, R. G., *Forests and Sea Power* (Archon Books, 1965).

Beer, J. J., *The Emergence of the German Dye Industry* (University of Illinois Press, 1959).

Black, George, *Lloyds Register of Shipping 1760–1960* (Lloyds Register of Shipping, 1960).

Clow, A., *The Chemical Revolution* (Batchworth Press, 1952).

Dickson, P. G. M., *Financial Revolution in England* (Macmillan, 1967).

Gerschenkron, A., *Bread and Democracy in Germany* (Fertig: New York, 1943).

Keeble, Frederick, *Fertiliser and Food Production* (Oxford University Press, 1932).

Lloyd, Christopher, *Lord Cochrane* (Longman, 1947).

Macintosh, George, *Biographical Memoirs of Charles Macintosh* (Blackie: Glasgow, 1847).

O'Dea, William, *A Social History of Lighting* (Routledge and Kegan Paul, 1958).

Read, J., Paine, C., Evans, J. C. and Todd, A. *Perkin Centenary, London* (Pergamon Press: Oxford, 1958).

CHAPTER 8

Bryan Donkin F. R. S. (Bryan Donkin Company Ltd, 1953).

Chandler, D. G., *The Campaigns of Napoleon* (Weidenfeld and Nicolson, 1967).

Clapperton, R. H., *The Papermaking Machine* (Pergamon Press: Oxford, 1967).

Critchell, J. T. and Raymond, J., *A History of the Frozen Meat Trade* (Constable, 1912).

Howard, M., *War in European History* (Oxford University Press, 1976).

Oman, C. W. C., *The Art of War in the Middle Ages* (Cornell University Press, 1885).

Ricketts, Howard, *Firearms* (Weidenfeld and Nicolson, 1964).

Taylor, F. L., *Art of War in Italy 1494–1529* (Cambridge University Press, 1921).

Vaughan, Richard, *Valois Burgundy* (Penguin: Harmondsworth, 1975).

Wedgwood, C. V., *The Thirty Years War* (Cape, 1938).

CHAPTER 9

Bowen, J. P., *British Lighthouses* (British Council/Longman, 1947).

Harvey, E. Newton, *A History of Luminescence* (American Philosophical Society: Philadelphia, 1957).

Hughes, Quentin, *Military Architecture* (Hugh Evelyn, 1974).

Josephson, Matthew, *Edison. A Biography* (McGraw Hill: New York, 1959).

Kiely, E. R., *Surveying Instruments* (Columbia University: New York, 1947).

Salmond, J. B., *Wade in Scotland* (Moray Press, 1934).

Tooley, R. V., *Maps and Mapmakers* (Batsford, 1949).

Warner, Philip, *The Medieval Castle* (Arthur Barker, 1971).

Webb, H. J., *Elizabethan Military Science* (University of Wisconsin, 1965).

Zworykin, V. K. and Morton, G. A., *Television* (Wiley: New York, 1940).

CHAPTER 10

Bell, D., *The Coming of the Post-Industrial Society* (Heinemann, 1974).

Index

Picture Acknowledgements

Ann Ronan Picture Library: 20, 31, 34 bottom, 35, 38, 39 right, 69, 75, 77, 88 bottom left/right, 121, 143, 147, 165, 196, 202, 208, 211, 237, 243, 258, 271, 273, 274. Ann Ronan Picture Library and E. P. Goldschmidt & Co. Ltd: 108 bottom, 255 bottom, 259. BASF Ludwigshafen: 204, 210. Bavaria Verlag/photo Emil Bauer: 107 top/bottom. Bayerische Staatsbibliothek: 33 top. Bettmann Archive: 113 bottom, 282 left. Biblioteca Estense, Modena/photo Scala: 97. Biblioteca Nazionale, Florence/photo Scala: 135. Bibliothèque de l'Arsenal/photo Giraudon: 60. Bibliothèque Municipale, Dijon: 91. Bibliothèque Nationale Service Photographique: 66, 67, 92, 96, 119, 184, 252. Bildarchiv Preussischer Kulturbesitz Kupferstichkabinett: 226, 227. Bodleian Library, Oxford: 22 left *MS.Marsh.144.-p43, 58 MS.Bod.968.f173r, 59 MS.Bod.264.f170v*, 64–5 *MS.Douce.88.f51*, 195 *MS.Douce.353.f31*. British Airways: 186. British Industrial Plastics: 216. Reproduced by permission of The British Library: 18 bottom, 23, 44, *MS. Harley 4205*, 49, 52, 53 bottom, 56, 57, 70, 90, 160, 255 top. Reproduced by courtesy of the Trustees of the British Museum/photo Michael Holford: 9, 194. Brown Brothers: 239, 282 right. Burgerbibliothek, Bern: 223. From Chapuis, A. and Gelis, E., *Le Monde des Automates*, Paris 1928: 108 top. Château Versailles/photo Giraudon: 228. Chevrolet: 155. Peter Clayton: 17. Clichés Musées Nationaux: 232. Mike Coles: 13, 83, 205. Copyright Bibliothèque Royal Albert 1er, Bruxelles: 131 *MS.IV.111.f13v*, 220 *MS.9.028.f6r*. Crown Copyright, Science Museum, London: 27, 32 bottom, 36–7, 41, 72, 73, 93, 144, 145, 161 bottom, 170, 171, 173, 177, 242, 269, 280 top. Crown Copyright, Victoria and Albert Museum, London: 53 top, 62–3, 109, 268. Director-General Meteorological Office: 37, 39 left. Drake Well Museum, Titusville: 180, 181. Emidata: 217. Fotomas Index: 64, 264. Dr Georg Gerster/John Hillelson Agency: 154. Guildhall Library, City of London: 199 (photo Cooper-Bridgeman Library), 265, 277 (photo Cooper-Bridgeman Library). Hamburger Kunsthalle: 161 top. Michael Holford: 12, 32 top, 51–1. Imperial War Museum, London: 247. Istanbul University Library/photo Erkin Emiroğlu: 114. David Jewitt: 281 top. From Jondet, M.G., *Atlas Historique de la Ville et des Ports d'Alexandrie*, Cairo 1921: 18 top. A. F. Kersting: 162, 163. Copyright © Dimitri Kessel, LIFE Picture Service: opp 1. Kunsthistorisches Museum, Vienna: 85, 224. Erich Lessing/Magnum from John Hillelson Agency: 8, 11. Macmillan London Ltd: 24, 25 bottom, 40 top, 42, 46, 47, 50 top, 55, 62 top, 63 top, 74, 78, 79, 86, 87, 94, 95, 99, 103, 110, 113 top, 122, 129, 130, 134, 136, 138–39, 141, 142, 143 left, 146, 147, 174, 175, 179 bottom, 182, 187, 208 top, 209, 212, 213, 222, 225, 229, 231, 233, 240, 244, 250, 256, 261, 276, 283 top, 284, 286. Macmillan London Ltd, courtesy Michealjohn Harris/BBC Television: 21, 29 top, 33 bottom, 178, 179, 272. Malaysian Rubber Research and Development Board (London): 203. Mansell Collection: 14, 16, 19, 26, 36, 60–1, 76, 88 top, 105, 112, 124, 193, 254, 275, 278. Mary Evans Picture Library: 34 top, 270. Mercedes-Benz (UK) Ltd: 183. Musée Condé, Chantilly/photo Giraudon: 214. Musées Royaux des Beaux-Arts de Belgique: 166–7. Museo Civico, Bologna/photo Scala: 80. Museo del Prado, Madrid: 158 (photo Michael Holford), 257 (photo Mas). Museo de Storia Veneziana/photo Scala: 190. Museum of the History of Science, Oxford University: 22 right, 30, 132. Museum of London: 201. NASA: 3, 248. National Maritime Museum, London/photo Michael Holford: 28 top/bottom, 189. Nederlansch Historisch Scheepvaart Museum, Amsterdam: 191. New York Public Library, Astor, Lenox and Tilden Foundations: 149. Osterreichische Nationalbibliothek: 29 bottom, 71. Patrimonio Nacional/photo Mas: 123. Photri: 4. Pierpont Morgan Library: 54. Plessey Microsystems Ltd: 117. Radio Times Hulton Picture Library: 176, 197, 230, 246, 280 bottom. Gerhard Reinhold, Leipzig-Mölkau: 159. Rijksmuseum, Amsterdam: 152. Reproduced by courtesy of The Royal Institution, London/photo Michael Holford: 245. Royal Scottish Geographical Society, Edinburgh: 260. St Bride Printing Library: 104. A. J. Samson: 40 bottom. Scala: 107, 218. Spartan Sheffield: 140. Staatsarchiv, Hamburg: 25. Stiftsbibliothek, St Gallen: 48. The Master and Fellows of Trinity College, Cambridge: 125. U.S. Air Force: 43. Yale University Art Gallery, Mabel Brady Garvan Collection: 150. Zentralbibliothek, Zürich: 126 *MS.C.54.ff34v, 35r,35v,36r*.